Nanostructures

Edited by Sadia Ameen,
M. Shaheer Akhtar and Hyung-Shik Shin

Published in London, United Kingdom

IntechOpen

Supporting open minds since 2005

Nanostructures
http://dx.doi.org/10.5772/intechopen.77456
Edited by Sadia Ameen, M. Shaheer Akhtar and Hyung-Shik Shin

Contributors
Ayi Ayi Anyama, Providence B. Ashishie, Bassey E. Inah, Emmanuel E. Khansi, Joseph O. Ogar, Chinyere A. Anyama, Hsin-Ming Cheng, Shoyebmohamad Shaikh, Rajaram Mane, Nanasaheb M. Shinde, Damin Lee, Abdullah M. Al-Enizi, Kwang Ho Kim, Chuan-Pei Lee, Yi-June Huang, Chun-Ting Li, Shiuan-Bai Ann, Han-Ting Chen, Parsoua Abedini Sohi, Mojtaba Kahrizi, Rajendra Sukhadeorao Dongre, Zahra Rafiei-Sarmazdeh, Seyed Morteza Zahedi, Sadia Ameen

First published in London, United Kingdom, 2020 by IntechOpen
IntechOpen is the global imprint of INTECHOPEN LIMITED, registered in England and Wales, registration number: 11086078, 7th floor, 10 Lower Thames Street, London, EC3R 6AF, United Kingdom
Printed in Croatia

British Library Cataloguing-in-Publication Data
A catalogue record for this book is available from the British Library

Additional hard and PDF copies can be obtained from orders@intechopen.com

Nanostructures
Edited by Sadia Ameen, M. Shaheer Akhtar and Hyung-Shik Shin
p. cm.
Print ISBN 978-1-78923-993-5
Online ISBN 978-1-78923-994-2
eBook (PDF) ISBN 978-1-83880-762-7

For EU product safety concerns:
IN TECH d.o.o., Prolaz Marije Krucifikse Kozulić 3, 51000 Rijeka, Croatia, info@intechopen.com or visit our website at intechopen.com.

Meet the editors

Professor Sadia Ameen obtained her PhD in Chemistry in 2008, and then moved to Jeonbuk National University. Presently she is working as an Assistant Professor at the Department of Bio-Convergence Science, Jeongeup Campus, Jeonbuk National University. Her current research focuses on dye-sensitized solar cells, perovskite solar cells, organic solar cells, sensors, catalyst, and optoelectronic devices. She specializes in manufacturing advanced energy materials and nanocomposites. She has achieved a Gold Medal in academics and the Merit Scholarship for the best academic performances. She is also the recipient of the Best Researcher Award. She has more than 120 peer-reviewed papers in the field of solar cells, catalyst, and sensors, and she has contributed to book chapters, edited books, and is an inventor/co-inventor of patents.

Professor M. Shaheer Akhtar completed his PhD in Chemical Engineering in 2008, from Jeonbuk National University, Republic of Korea. Presently, he is working as an Associate Professor at Jeonbuk National University, Republic of Korea. His research interests are the photo-electrochemical characterizations of thin film semiconductor nanomaterials, composite materials, polymer-based solid-state films, solid polymer electrolytes and electrode materials for dye-sensitized solar cells (DSSCs), hybrid organic-inorganic solar cells, small molecules based organic solar cells, and photocatalytic reactions.

Professor Hyung-Shik Shin received his PhD in the kinetics of initial oxidation Al (111) surface from the Cornell University, USA, in 1984. He is a Professor in the School of Chemical Engineering, Jeonbuk National University and also President of the Korea Basic Science Institute (KBSI), Gwahak-ro, Yuseong-gu, Daejon, Republic of Korea. He has been a promising researcher and has visited several worldwide universities as a visiting professor/invited speaker. He is an active executive member of various renowned scientific committees such as KiChE, copyright protection, KAERI, etc. He has extensive experience in electrochemistry, renewable energy sources, solar cells, organic solar cells, charge transport properties of organic semiconductors, inorganic-organic solar cells, biosensors, chemical sensors, nano-patterning of thin film materials, and photocatalytic degradation.

Contents

Preface

In today's modern world, nanostructure materials can be designed and grown successfully in the laboratory with the advent of new synthesis methods. Years of scientific efforts have shown that nanostructure materials are rising as the modern tools of technology and development. Nanostructure materials are one such exceptional class of modern materials that have attracted tremendous curiosity among researchers around the world due to their potential in technology. A nanostructure is a structure of intermediate size between microscopic and molecular structures. In describing nanostructures, it is necessary to differentiate between the numbers of dimensions in the volume of an object, which are on the nanoscale. Additionally, nanostructure materials, by virtue of their unique physical and chemical properties such as thermal stability, chemical resistance, hardness, large surface area and porosity, electron transfer capacity, conductance and impedance, have become remarkably useful in a wide range of applications. In short, nanostructure materials are worthy of being applied practically and efficiently as smart functional materials in several developmental areas including but not limited to industrial process-control, environmental remedies, semiconductors, and energy devices.

This book is designed to highlight the importance, functionality, and applicability of nanostructure materials. The chapters in this book offer clear information on the recent advances in the synthesis, characterization, and application of these materials to energy conversion such as perovskite solar cells, dye sensitized solar cells, and sensors. We hope this book will help describe the current position of nanostructure materials in the technological sphere as well as encourage scientists and engineers in deeper exploration of these materials to boost the technological advancement. Written by an outstanding group of experts in the field, this book presents an exciting and fresh compilation of the latest advances and developments in nanostructure materials. As a result, the highly interdisciplinary nature of this book would be of profound interest for a broad audience, including academic and industrial researchers. It can also be used in the classroom for graduate students focusing on chemical engineering, surface and interface science, adsorption science and technology, zeolite science, and nanostructure materials science. We believe that this book will also be valuable to many entrepreneurial and business people who are in the process of trying to better understand and valuate nanotechnology and novel nanostructure materials for future high-tech emerging applications and disrupting industries. To keep the length of the book manageable, we have limited the discussion to nanostructure materials; however, we have tried to encompass as many related details as possible. We would like to thank all the authors who contributed to this book with their knowledge, efforts, and time. We are also grateful to the copyright owners, and IntechOpen for their support in publishing and making this book a reality.

Sadia Ameen
Advanced Materials and Devices Laboratory,
Department of Bio-Convergence Science,
Jeongeup Campus,
Jeonbuk National University,
Republic of Korea

M. Shaheer Akhtar
New and Renewable Energy Material Development Center (NewREC),
Jeonbuk National University,
Republic of Korea

Hyung-Shik Shin
School of Chemical Engineering,
Jeonbuk National University,
Republic of Korea

Korea Basic Science Institute (KBSI),
Gwahak-ro, Yuseong-gu, Daejon, Republic of Korea

Chapter 1

Introductory Chapter: Prospects of Nanostructured Materials

Sadia Ameen

1. Introduction

In the past few decades, there has been a worldwide surge of research interests in the field of nanoscience and nanotechnology. It has been realized that it has an impeccable potential to revolutionize the modern technology in countless ways. The field of "nanoscience" has attracted much attention from scientists and engineers due to its interesting scientific aspects, and the materials thus formed have novel physicochemical properties. In the beginning, the "miniaturization" of new devices and systems being the central theme of modern technologies was considered, but these days the subject of "nanomaterials" indeed happens to be very demanding. It has overwhelming contribution in technological advancement which has already started to have a global commercial impact.

In scientific terms, nanomaterials are basically any substance having a size of the order of 10^{-9} m, i.e., a billionth of a meter, and they are generally considered as "nanomaterials" when their sizes are typically lesser than 100 nm. In other words, nano-sized materials are the "collection of distinguishable units," each of which is made up of a limited number of atoms. These units have specific shape and crystallinity and have at least one of its dimensions of about a few nanometers. A few common examples of naturally occurring nanomaterials are particulate matter from volcanic eruptions, different types of bacteria, cosmic dust, protrusions on lotus leaf that limits its wettability, etc. Also there are numerous man-made nanomaterials which include but are not limited to fullerenes, graphene, single-/multi-walled carbon nanotubes (CNTs), silver nanoparticles, metal oxide nanorods, quantum dots, etc. These nanomaterials have fascinating properties and, hence, are technologically important. For instance, due to the appropriate magnitudes of tensile strength, stiffness, and conductivity, graphene or CNTs have potential applications in catalysis, sensors, and drug deliveries. Nano-sized silver has displayed its capability of oxygen inhibition which could be utilized to pacify the burn injuries of living organisms. On the other hand, quantum dots of semiconducting materials are best suited in display and device technologies related to LED, memory storage, and solar cells [1–3]. In this manner, a variety of nanomaterials are potentially applicable in every sphere of modern technology which includes automotive, aerospace, cutting and device manufacturing industries, printing and color imaging, armor, computer chips, and biomedical and pharmaceutical fields as well.

The fact that attributes a special character to nanomaterials and variation in their overall behavior is their significantly small size. This generates "statistical size effect" that plays a vital role in altering most of the physical and chemical properties of these materials. Due to the reduction of the size to very small dimensions, typically 1–100 nm, more and more atoms are exposed to the surroundings and result into a vast increment of surface area to volume ratio for that material. This material property is particularly responsible for assigning high specific surface area to the material, which remarkably improves many processes such as adsorption, catalysis, charge storage, and transfer on the surface of the material and hence opens avenues for its applications in sensors, supercapacitors, and batteries as well. Another feature associated to the reduction of particle size down to nano-regime is "quantum confinement effect." The valence and free electrons of atoms in these nano-sized assemblies undergo considerable variations in their energy states as compared to their features in the macroscopic state. This affects the electronic band structure of the material in such a way that the separations between the continuous energy bands or the individual electronic energy levels alter. Subsequently the occupation and transition of electrons in these levels also change substantially which results into variation in the electronic, magnetic, and optical properties of that material [4]. There are numerous such examples wherein physical/chemical properties have been remarkably improved or modified through controlling the sizes and shapes of nanostructures. These interesting variations highlight the need of nanomaterials in newer technologies, and hence the designing and development of nanomaterials with tailor-made properties fitting specific applications happens to be the current goal of researchers working in this field. Efforts are also being taken to synthesize nanomaterials in different shapes such as nanorods, nanoparticles, nanobelts, nanotubes, and nanoshells with variations in their compositions as well.

Efficient synthesis of nanomaterials has always been a worldwide challenge. Numerous methods have already been devised and are also evolving continuously to fulfill this need. Methods such as auto-combustion, hydrothermal, coprecipitation, and sol–gel can be categorized as chemical methods to synthesize nanomaterials [5–8], whereas the methods such as solid-state route and high-energy mechanical milling can be termed as the physical methods. Also there are some more sophisticated techniques such as thermal evaporation, chemical vapor deposition, sputtering, pulsed laser ablation, pyrolysis, spin coating, etc. for the synthesis of high-quality thin films having nanocrystalline properties [9–11]. Every technique has its own merits and peculiarities for the synthesized nanomaterial. The solid-state methods are simple to yield the final product but may require high temperatures for the formation of desired compound. On the other hand, the deposition techniques are far more sophisticated than the other methods and hence become expensive; however, the synthesis can be controlled multi-parametrically which yield nanocrystalline thin films of superior quality. Using these techniques, variation in most of the properties such as morphology, composition, grain size, shape, thickness, crystal structure, and even non-equilibrium phases of synthesized materials could be achieved. With the advent of highly sophisticated analytical equipment such as atomic force microscopy (AFM) and scanning tunneling microscopy (STM), field emission scanning electron microscopy (FESEM), transmission electron microscopy (TEM), X-ray diffraction (XRD), and X-ray photoelectron spectroscopy (XPS), it is possible to observe and study these nano-sized materials from the point of view of future technological applications.

In the context of synthesizing nanomaterials, recently Ameen et al. discovered a fast and highly effective growth method for thin film of nanocrystalline tungsten oxide (WO_3) through one-step hot filament chemical vapor deposition (HFCVD) method at low temperature. The grown monoclinic WO_3 crystal structure exhibited the morphology of cauliflower-like nanostructure (WCNs). The grown WO_3 thin film was applied for the detection of hydrogen (H_2) gas at 100 ppm [11]. From the H_2 sensing behavior of WCN thin film, the appreciable sensitivity of the WCN thin film was obtained only after ~200°C and reached a high magnitude of ~87% at optimum operating temperature of ~250°C, as shown in **Figure 1**.

In year 2019, another work by Ameen and co-workers was reported on the synthesis of hexagonal zinc oxide (ZnO) nanopyramids (NPys) for the application of electrochemical sensor [12]. The synthesized hexagonal pyramid shape structures possessed the average size of ~2–3 μm with the base length of ~2 μm and a top length of ~600 nm. From **Figure 2**, it is seen that the agglomeration of several hexagonal disks formed pyramid-like structures.

Nonenzymatic glucose biosensor was developed by utilizing the electrochemically grown nanocages-augmented polyaniline nanowires (NCa-PANI NWs) on silicon (Si) substrate [13]. Figure exhibits that the grown NCa-PANI NWs are distributed uniformly on the entire surface of silicon (Si) substrate, confirming the growth of highly dense NCa-PANI NW structures (**Figure 3**).

Kim et al. designed field effect transistor using iron-nickel co-doped ZnO nanoparticles, $Zn_{0.97}Fe_{0.01}Ni_{0.02}O$ NPs, for the electrochemical detection of hexahydropyridine chemical, as shown in **Figure 4** [14]. $Zn_{0.97}Fe_{0.01}Ni_{0.02}O$ NP-modified FET sensor showed the high sensitivity of ~62.28 μA μM^{-1} cm^{-2} and a good detection limit of ~79 μM with correlation coefficient (R) of ~0.96405 and a short response time (10 s) towards hexahydropyridine chemical.

Figure 5 shows the synthesized ZnO nanoflowers (NFs) for the photocatalytic degradation of bromophenol (Bph) dye. ZnO NFs as catalysts displayed a rapid degradation of Bph-dye with the degradation rate of ~96% within 120 min under the UV light irradiation [15].

Jang and co-workers synthesized porous cobalt oxide (Co_3O_4) nanocubes (NCs), as shown in **Figure 6**, and investigated the capacitive properties of porous Co_3O_4

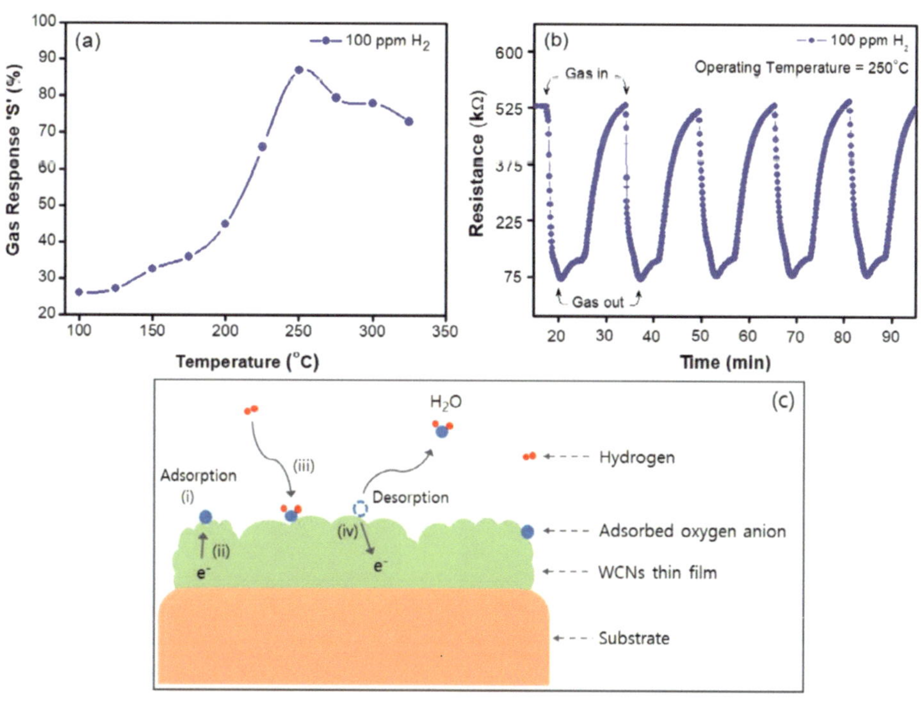

Figure 1.
(a) Gas sensing response of grown WO_3 thin film to H_2 gas at various temperatures, (b) response recovery profile with dynamic repeatability at ~250°C for 100 ppm H_2, and (c) illustration of H_2 sensing mechanism on grown WO_3 thin film. Reproduced from reference [11].

Figure 2.
Low (a) and high (b) magnification FESEM images of hexagonal ZnO NPys. Reproduced from ref. [12].

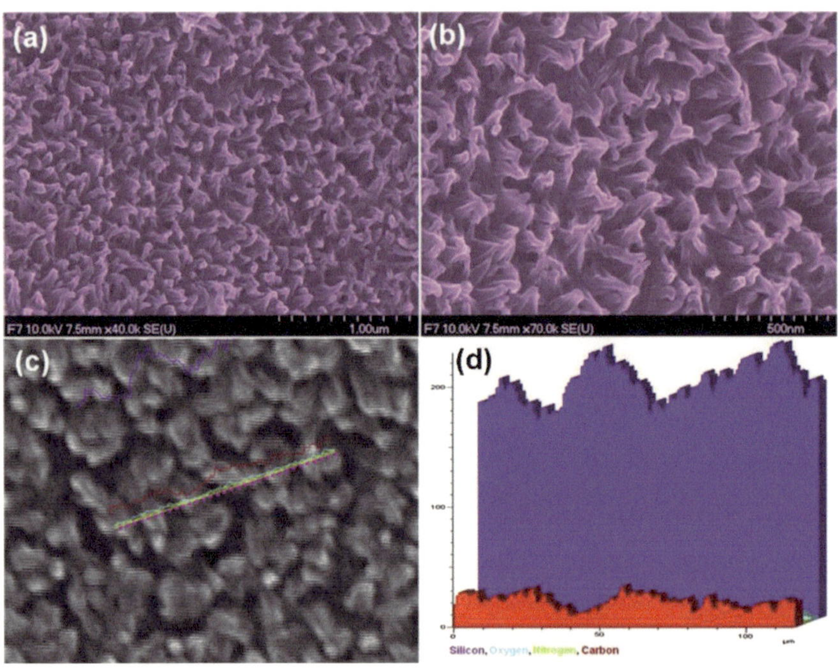

Figure 3.
Low (a) and high (b) magnification FESEM images, element line scanning mapping (c), and its corresponding profile (d) of PANI NW electrode. Reproduced from ref. [13].

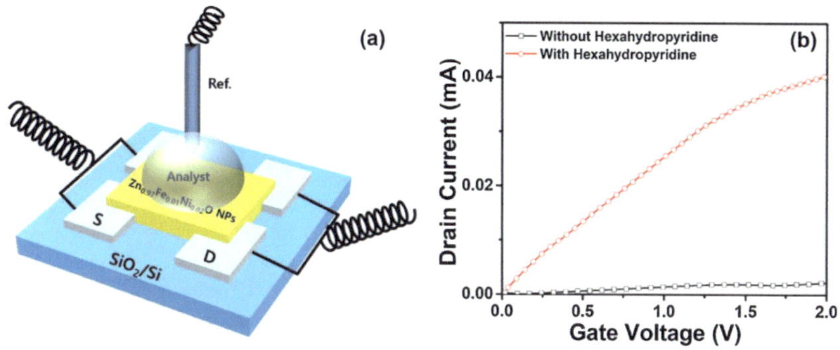

Figure 4.
(a) A diagram represented the fabricated FET sensor with $Zn_{0.97}Fe_{0.01}Ni_{0.02}O$ NPs and (b) I_d-V_g response of FET sensor without and with hexahydropyridine (100 nM) in 0.01 M PBS. Reproduced from ref. [14].

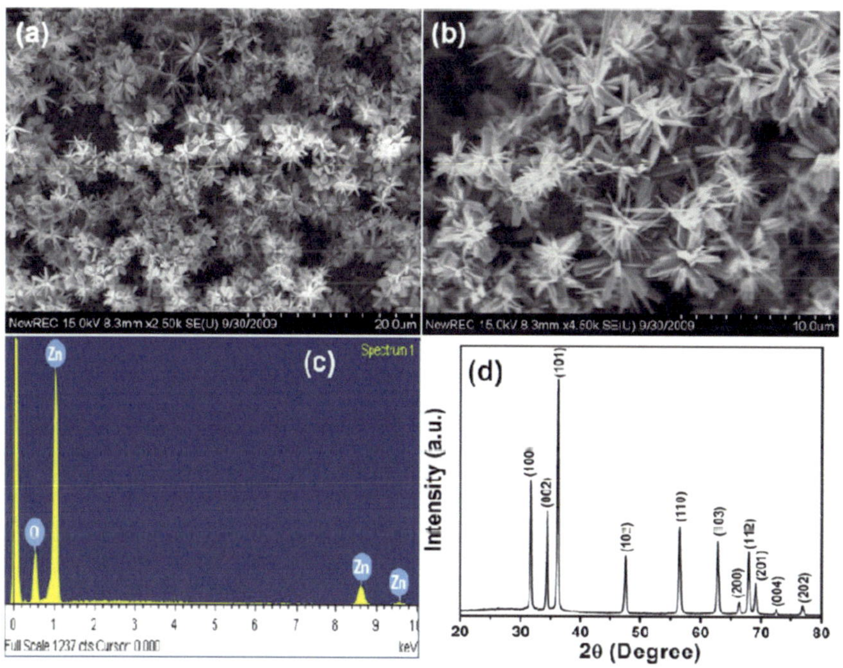

Figure 5.
(a) Low and (b) high magnification FESEM images, (c) EDX spectrum, and (d) XRD analysis of ZnO NFs. Reproduced from ref. [15].

NCs electrode by cyclic voltammetry (CV) in 6 M KOH electrolyte. High specific capacitance of ~430.6 F/g at a scan rate of ~10 mV s^{-1} was observed. The capacity retention of up to ~85% after 1000 cycles was shown by the fabricated porous Co_3O_4 NCs electrode. The porous Co_3O_4 NCs showed excellent structural stability through cycling with promising capacity retention which suggested a good quality of porous Co_3O_4 NCs as electrochemical supercapacitor electrode [16].

This prologue provides a glimpse of nanomaterials and the way their properties change and also gives an idea about their potent applicability which can shift the entire technological paradigm. This book has been designed to induce scientific interest in the minds of young researchers and engineers to undertake research in this field and work across the ever-expanding boundaries of nanoscience and nanotechnology.

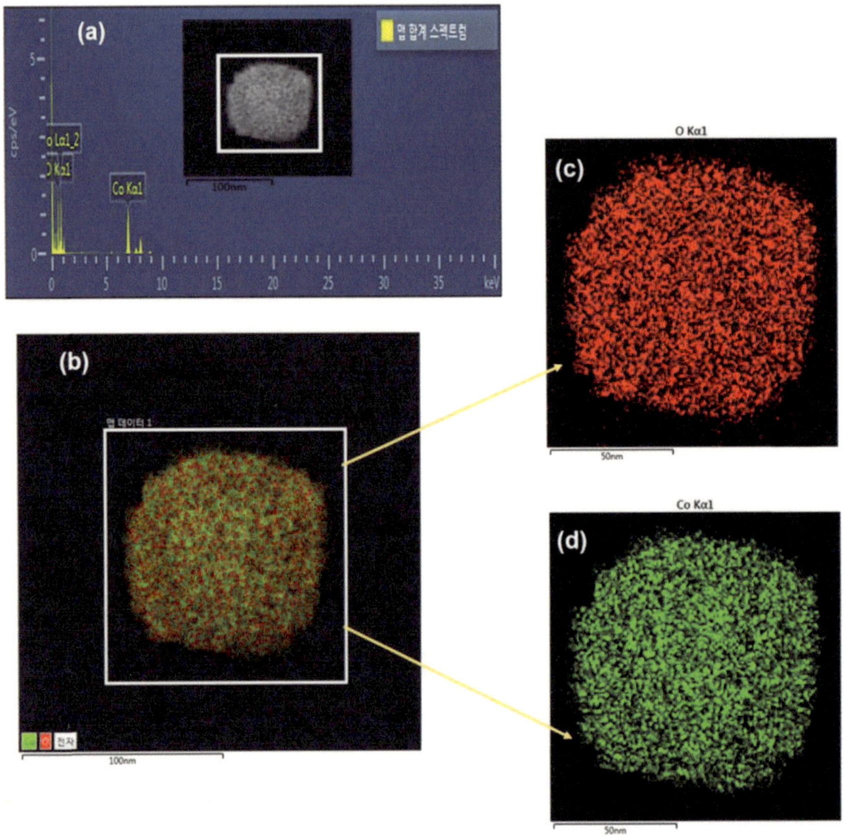

Figure 6.
The representation of (a) TEM-EDX including profile and (b–d) elemental mapping images of porous Co_3O_4 NCs. Reproduced from ref. [16].

Author details

Sadia Ameen
Advanced Materials and Devices Laboratory, Department of Bio-Convergence Science, Jeonbuk National University, Republic of Korea

*Address all correspondence to: sadiaameen@jbnu.ac.kr

References

[1] Ameen S, Seo HK, Akhtar MS, Shin HS. Novel graphene/polyaniline nanocomposites and its photocatalytic activity toward the degradation of rose Bengal dye. Chemical Engineering Journal. 2012;**210**:220-228

[2] Schmidt G. Nanoparticles. Weinheim, Germany: WILEY-VCH Verlag GmbH & Co. KGaA; 2004

[3] Lewis LN. Chemical catalysis by colloids and clusters. Chemical Reviews. 1993;**93**:2693-2730

[4] Godbole RV. Thesis Repository. India: Jayakar Knowledge Resource Centre, Savitribai Phule Pune University; 2018. Available from: lib.unipune.ac.in:8080/xmlui/handle/123456789/8621

[5] Tao S, Gao F, Liu X, Sorensen OT. Preparation and gas-sensing properties of $CuFe_2O_4$ at reduced temperature. Materials Science and Engineering B. 2000;**77**:172-176

[6] Kolahalam LA, Viswanath IVK, Diwakar BS, Govindh B, Reddy V, Murthy YLN. Materials Today: Proceedings. 2019;**18**:2182-2190

[7] Gadkari AB, Shinde TJ, Vasambekar PN. Structural analysis of Y3+-doped Mg–Cd ferrites prepared by oxalate co-precipitation method. Materials Chemistry and Physics. 2009;**114**:505-510

[8] Ameen S, Akhtar MS, Seo HK, Shin HS. Advanced ZnO–graphene oxide nanohybrid and its photocatalytic Applications. Materials Letters. 2013;**100**:261-265

[9] Ameen S, Akhtar MS, Shin HS. Hydrazine chemical sensing by modified electrode based on in situ electrochemically synthesized polyaniline/graphene composite thin film. Sensors and Actuators B: Chemical. 2012;**173**:177-183

[10] Anderson A. Raman Study of Ceramic Tungsten Trioxide at Low Temperatures. Spectroscopy Letters. 1976;**9**:809-819

[11] Godbole R, Ameen S, Nakate UT, Akhtar MS, Shin HS. Low temperature HFCVD synthesis of tungsten oxide thin film for high response hydrogen gas sensor application. Materials Letters. 2019;**254**:398-401

[12] Kim EB, Abdullah S, Ameen MS, Akhtar HS, Shin J. Environment-friendly and highly sensitive dichloromethane chemical sensor fabricated with ZnO nanopyramids-modified electrode. Journal of the Taiwan Institute of Chemical Engineers. 2019;**102**:143-152

[13] Ameen S, Akhtar MS, Shin HS. Nanocages-augmented aligned polyaniline nanowires as unique platform for electrochemical non-enzymatic glucose biosensor. Applied Catalysis A: General. 2016;**517**:21-29

[14] Kim EB, Ameen S, Akhtar MS, Shin HS. Iron-nickel co-doped ZnO nanoparticles as scaffold for field effect transistor sensor: Application in electrochemical detection of hexahydropyridine chemical. Sensors and Actuators B: Chemical. 2018;**275**:422-431

[15] Ameen S, Akhtar MS, Shin HS. Speedy photocatalytic degradation of bromophenol dye over ZnO nanoflowers. Materials Letters. 2017;**209**:150-154

[16] Jang GS, Ameen S, Akhtar MS, Shin HS. Cobalt oxide nanocubes as electrode material for the performance evaluation of electrochemical supercapacitor. Ceramics International. 2018;**44**:588-595

Chapter 2

Nano/Micro-Structured Materials: Synthesis, Morphology and Applications

Ayi A. Ayi, Providence B. Ashishie, Emmanuel E. Khansi, Joseph O. Ogar, Chinyere A. Anyama and Bassey E. Inah

Abstract

Materials with structural elements, clusters and crystallites or molecules with size dimension in the range 1–100 nm and/or 4–20 Å have found potential and real applications as antimicrobial agents, catalysts, nano-filters in waste water treatments and scale forming ions removal etc. These nano/micro-structured materials possess large surface area which is one of the most important properties needed in different fields of applications. In this short review, the different protocols available for the synthesis ranging from green chemistry to chemical reduction methods, structural characterization, morphology and applications of nanostructured materials such as layered double hydroxides, silver and molybdenum oxides have been discussed.

Keywords: nanomaterials, chemical reduction, green chemistry, layered double hydroxides, antimicrobial agents

1. Introduction

The synthesis of inorganic nanostructured materials, their stabilization, morphology and properties are the main issues of interest in different research groups. The interest is not only for their applications in chemical technology, catalysis, magnetic data storage and sensing [1], but also for their antimicrobial activities and waste-water treatment. More often than not, marine sediments and water environments are contaminated by urban runoffs, industrial and domestic effluents and oil spills [2]. The presence of such contaminants as polycyclic aromatic hydrocarbons (PAHs), poly chlorinated biphenyls (PCBs), dichlorodiphenyltrichloroethane (DDT) and heavy metals like mercury, lead and manganese poses risk to both human health and the aquatic biota [3]. Functionalized magnetic metal oxide nanoparticles [4, 5], silver nanoparticles as well as nanostructured layered double hydroxides have proven useful in the remediation of toxic waste in the water environment, inhibition of bacterial and fungi activities [6] and removal of scale forming ions from oil wells [7].

Different protocols have been employed in the synthesis of nanostructured materials by different research groups. These synthetic protocols are generally grouped into two categories: "top-down" and "bottom-up" approach. In the top-down approach (milling or attrition), bulk solid is broken into smaller and smaller portions, until nanometer size is reached. The bottom-up method of nanoparticle synthesis involves nucleation of atoms followed by their growth in a self-assembly fashion to form the nanomaterial. Here we will look at the different nanostructured materials prepared by bottom-up methods such as hydrothermal, [8, 9] combustion synthesis [10], gas-phase methods [11, 12], microwave synthesis and sol-gel processing [13].

2. Nano/micro-structured materials

2.1 Nanostructured layered double hydroxides

Layered double hydroxides (LDHs) with the hydrotalcite-like structure are materials attracting interest to many research groups due to their use in many different fields, such as anion scavengers, catalysts, catalyst precursors, hosts for drugs controlled delivery, contaminant and radionuclide removal from acidic mine pit water etc. [14–16]. Layered double hydroxides (LDHs) are a class of anionic

Figure 1.
(a) SEM micrograph of Mg-Al(OH)$_2$PO$_4$, 1; (b) SEM micrograph of Mg-Al(OH)$_2$PO$_4$PF$_6$, 2; (c) SEM micrograph of Ca-Al(OH)$_2$SO$_4$, 3; (d) SEM micrograph of Ca-Al(OH)$_2$PO$_4$PF$_6$, 4. Source: Khansi et al. [7].

clays with the structure based on brucite ($Mg(OH)_2$)-like layers [17–21]. The lattice structure of LDHs with the general formula $[M_{2+1}{-}xM^{3+}_{x}(OH)_2]^{x+}(A_n^{-})_{x/n}\cdot yH_2O$, have a positively charged brucite-shaped layers, consisting of a divalent metal ion M^{2+} (e.g., Ca^{2+}, Zn^{2+}, Mg^{2+}, and Ni^{2+}) octahedrally surrounded by six OH^- hydroxyl groups [22–24]. The substitution of the M^{2+} metal with a trivalent M^{3+} cation gives rise to the periodic repetition of positively charged sheets (lamellas) alternating with charge-counter balancing A_n^- ions. According to Kovačević et al. [25], synthetic layered double hydroxides (LDHs) have been used as adsorbents and catalyst supports. The surface area and pore volume of LDHs can be increased by pillaring the interlayer with large organic and inorganic anions [19, 26, 27]. The combination of LDHs with negatively charged polymers can result in the formation of a hybrid layer component for the preparation of nano-hybrid thin films [28–30].

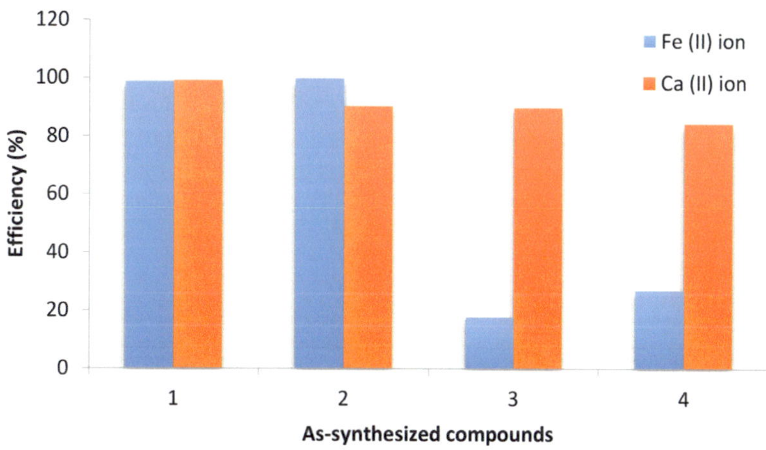

Figure 2.
Results for the removal of Fe^{2+} and Ca^{2+} ions from aqueous solutions by compounds 1–4. Source: Khansi et al. [7].

Figure 3.
(a) SEM micrograph of UCIM-1; (b) SEM micrograph of UCIM-2; (c) SEM micrograph of UCIM-3; (d) SEM micrograph of UCIM-4; (e) SEM micrograph of UCIM-5. Source: Ogar [37].

When LDHs which are formed by incorporating both divalent and trivalent metal ions into the octahedral lattice, are calcined at higher temperature, mixed metal oxide nanomaterials are formed with improved catalytic activity [31–33]. LDHs are also used as supports in bio-nanocomposites for medical applications [34].

Khansi et al. [7] synthesized four metal-aluminum layered double hydroxides (LDHs): Mg-Al$(OH)_2PO_4$, **1**, Mg-Al$(OH)_2$ PO_4 PF_6, **2**, Ca-Al$(OH)_2SO_4$, **3** and Ca-Al$(OH)_2PO_4PF_6$, **4**, by co-precipitation method followed by mild hydrothermal processing at 60°C. In **Figure 1**, we present the SEM micrographs of the as-synthesized LDHs. The synthesized nano-/microstructured LDHs of Mg-Al$(OH)_2PO_4$, **1** (**Figure 1a**), [Mg-Al$(OH)_2PO_4PF_6$], **2** (**Figure 1b**) and [Ca-Al$(OH)_2PO_4PF_6$] **4** (**Figure 1d**) have their Mg—O—Al—OH and Ca—O—Al—OH layers intercalated with PO_4^{3-} and/or $[PO_4PF_6]^{4-}$ anions. Their micrographs consists of uniformed nano/microspheres of M—Al layered double hydroxides, while the Ca—O—Al—OH layer intercalated with SO_4^{2-} anions (**Figure 1c**) consisted of hexagonal nano-/microplates.

The results of the column adsorption studies showed that there is significant potential for using the synthesized nanostructured LDHs as nano filters in removing ions responsible for scale formation in oil wells. According to their report, compounds **1** and **2** removed Fe^{2+} with greater efficiency, while all the synthesized LDHs nanostructures effectively removed Ca^{2+} from the oil wells as shown in **Figure 2**.

2.2 Nanostructured metal phosphonates

A surfactant-assisted synthesis route to prepare nanometer-sized metal phosphonate particles for use as an inhibitor in porous media for scale control has been reported [35, 36]. In their method, aqueous solution of calcium chloride and zinc chloride were mixed with basic solution containing phosphonate scale inhibitors such as diethylenetriamine-penta(methylene phosphonic acid) (DTPMP) in the presence of sodium dodecyl sulfate as surfactant. Following similar procedure, Ogar [37] used hydrothermal technique in the preparation of five nanostructured metal phosphonates. In a typical synthesis, 1.5 g of sodium dodecyl sulfate (SDS) was added to 40 cm^3 of 0.25 M solution of the $CaCl_2$. This was accompanied by constant stirring for 20 min (pH = 6). Then 25 cm^3 of 1.0 M solution of phenylphosphonic acid (pH = 2) was added dropwise to the above solution with constant stirring. The mixture (pH = 3) was allowed to stir for 1 h after which it was transferred into ACE glass tubes and heated for 8 h at a temperature of 120°C. This was followed by centrifugation at 2000 RPM for 20 min, filtration and drying of the nanomaterials in an electric incubator at 80°C. The product obtained was designated UCIM-1 (University of Calabar Inorganic Material). Similar procedure was employed in the synthesis of UCIM-2, UCIM-3, UCIM-4 and UCIM-5 with different complexing agent as aminomethylphosphonic acid (AmMePhA), 4,4′-biphenyl-bis-phosphonic acid (BBPhA) and different metal salt such as zinc acetate.

The SEM micrographs of UCIM-1 and UCIM-2 showed clusters of micro-rods (**Figure 3a** and **b**). The zinc phosphonates prepared in the presence of surfactant, UCIM-3 consisted of particles in the size range of 30–100 nm (**Figure 3c**), whereas UCIM-4 and UCIM-5 prepared without the surfactant consists of monodispersed nanoballs (**Figure 3d** and **e**). The synthesized nanostructured zinc and calcium phosphonates were applied in the removal of scale-forming ions from aqueous solution. A similar procedure reported for LDHs [7] was used. Small amount of glass wool was first introduced into the syringe followed by 0.2 g of the nanoparticles. Two molar solution of $CaCl_2$ was then introduced. The set-up was left for 12 h during which the solution eluted was collected in sample bottles and analyzed using AAS. This was repeated for $FeSO_4.7H_2O$. The removal efficiency of Fe^{2+} and Ca^{2+} by metal phosphonate nanoparticles (UCIM-n) in percentage was calculated using the formula: efficiency (%) = C_e/C_o (C_o = initial concentration of the ions, C_e = eluded concentration of the ions after passing through the packed column).

The removal efficiency of scale forming ions in oil wells by the synthesized metal phosphonate nanoparticles was studied with the help of atomic absorption spectroscopic technique. The removal efficiency of Ca^{2+} and Fe^{2+} ions are presented in **Tables 1** and **2** respectively. The results showed that the surfactant-assisted zinc nanostructured material (UCIM-4) removed up to 92.04% of Fe^{2+} compared to UCIM-5 prepared without surfactant, which showed only 75.88% removal. The compound UCIM-4 adsorbed Ca^{2+} the most, followed by surfactant-assisted UCIM-1 Ca-nanoparticles with a very high percentage adsorption of 94.29 as presented in **Figure 4**.

2.3 Metal/bimetallic/metal-oxide nanoparticles

There are a number of interesting reports on the use of ionic liquids as a solvent and/or stabilizing agent in the synthesis of metal/metal-oxide nanoparticles. The interest is not only due to the fact that ILs can form extended hydrogen-bond networks at the liquid state, but also for their nanostructural organization which is being used as a driver for spontaneous formation of nanostructured materials [38] Taubert and co-workers [39] reported on the effects of a set of ILs based on the 1-ethyl-3-methylimidazolium cation and different anions on the formation of gold nanoparticles. They demonstrated that the nature of the ionic liquid anion was crucial in the synthesis of the gold nanoparticles. For instance, anions such as methanesulfonate (MS) and trifluoromethanesulfonate (TfO) stabilized the formation of particles in the size range of 5–7 nm at low temperature, whereas with ethyl

Nanoparticles	Initial amount (ppm)	Amount eluted (ppm)	Amount adsorbed (ppm)	Percentage adsorption
UCIM-1	7.7780	0.4438	7.3342	94.29
UCIM-2	7.7780	1.1512	6.6268	85.20
UCIM-3	7.7780	1.0251	6.7529	86.82
UCIM-4	7.7780	0.5729	7.2051	92.63
UCIM-5	7.7780	0.2399	7.5381	96.92

Table 1.
Removal efficiency of Ca^{2+} by the synthesized metal phosphonate nanoparticles (source: Ogar [37]).

Nanoparticles	Initial amount (ppm)	Amount eluted (ppm)	Amount adsorbed (ppm)	Percentage adsorption
UCIM-1	9.8526	4.1779	5.6747	57.60
UCIM-2	9.8526	5.1040	4.7486	45.46
UCIM-3	9.8526	5.3979	4.4547	45.21
UCIM-4	9.8526	0.7843	9.0683	92.04
UCIM-5	9.8526	0.2399	7.5381	96.92

Table 2.
Removal efficiency of Fe^{2+} by the synthesized metal phosphonate nanoparticles (source: Ogar [37]).

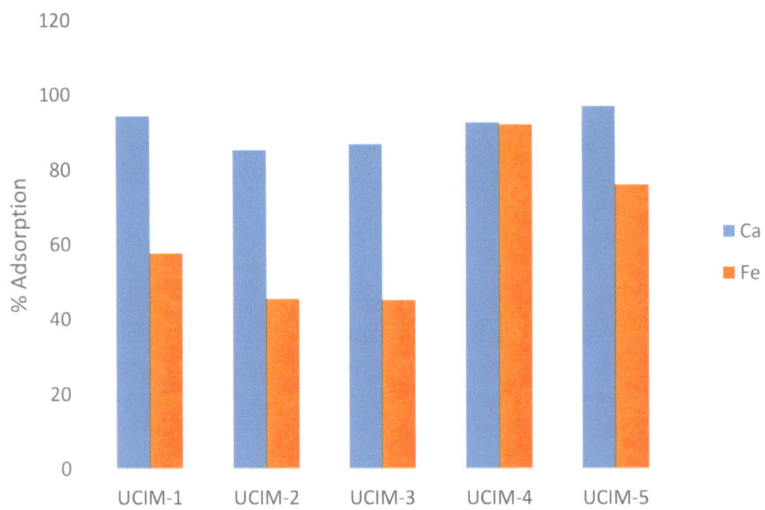

Figure 4.
Bar chart showing percentage adsorption of Ca^{2+} and Fe^{2+} by the different metal phosphonate nanoparticles. Source: Ogar [37].

sulfate (ES), polydispersed particles in the size range 15–20 nm were formed at all reaction temperature. In their report, they concluded that specific IL-gold interactions were responsible for the formation of gold particles with an IL-specific shape, size, and aggregation behavior.

Ayi et al. reported on a wet chemical approach in synthesizing titanium nanoparticles in imidazolium-based ionic liquids (ILs) under reducing conditions [40]. It was established that nanosize particles were formed with some ILs being adsorbed on the nanoparticles, thus providing the needed stabilization.

In continuation of the work on anion effects on nanoparticles formation, stable Mo and molybdenum oxide nano- and microparticles were synthesized in ILs with a variety of anions under reducing conditions by Ayi and co-workers [41]. XRD and TEM revealed a strong influence of the IL anion on the particle sizes, shapes, and crystal structures. The influence of the IL cation and the reaction temperature was found to be much less pronounced.

Very recently, Ayi's research group has used plant extracts (*Kigelia africana* fruits) to prepare silver nanoparticles and copper-silver bimetallic nanostructures and their antimicrobial potentials evaluated [42]. The aqueous extract of *Kigelia africana* was utilized in the synthesis of both metallic and bimetallic nanoparticles without additional stabilizing agent. The investigation revealed that average particle size of 10 nm were formed and stabilized by the active components of the plant extract, which were adsorbed on the surface of the particles. The synthesized particles were found to inhibit the growth of both Gram-negative and Gram-positive bacteria more than any of antibiotics tested in the study. The bimetallic nanoparticles demonstrated effectiveness against *S. aureus* with maximum ZOI of 27 mm. Walter et al., made use of 1-ethyl-3-methylimidazolium acetate (EMIMOAC) and 1-buthyl-3-methylimidazolium acetate (BMIMOAC) ionic liquids to prepare cobalt nanoparticles (CoEMIMOAC or CoBMIMOAC) via chemical reduction method [43]. The SEM and UV-Vis techniques confirmed the formation of nanoparticles (**Figure 5**). There was no difference between the nanocrystals formed in EMIMOAC or BMIMOAC indicating that the cationic part of the ionic liquid did not play significant role during nucleation and growth of the particles. The inhibition zone diameter (IZD) showed moderate susceptibility of the *Staphylococcus aureus* to CoBMIMOAC (15 mm) but high susceptibility to

CoEMIMOAC with a larger inhibition zone diameter of 23 mm. Gram-negative microorganism *E. coli* is moderately susceptible to both CoEMIMOAC (14 mm), and CoBMIMOAC (15 mm) (**Figure 6**).

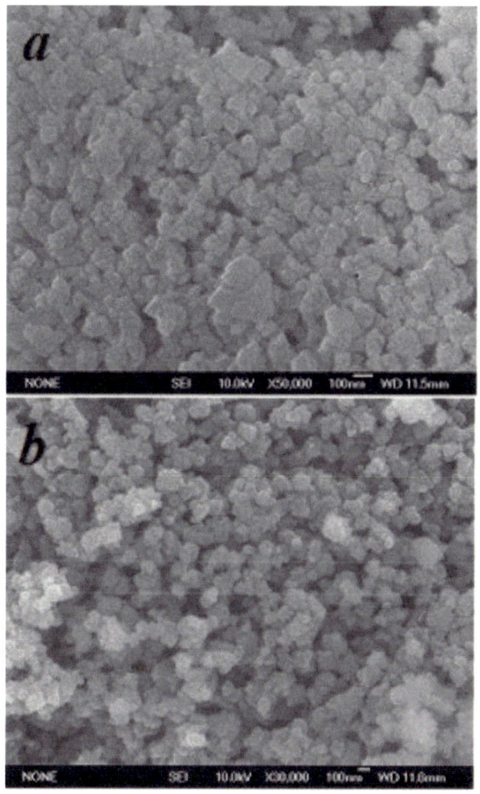

Figure 5.
(a) SEM micrograph of cobalt nanocrystals prepared in EMIMOAC. Source: John et al. [43]; (b) SEM micrograph of cobalt nanocrystals prepared in BMIMOAC. Source: John et al. [43].

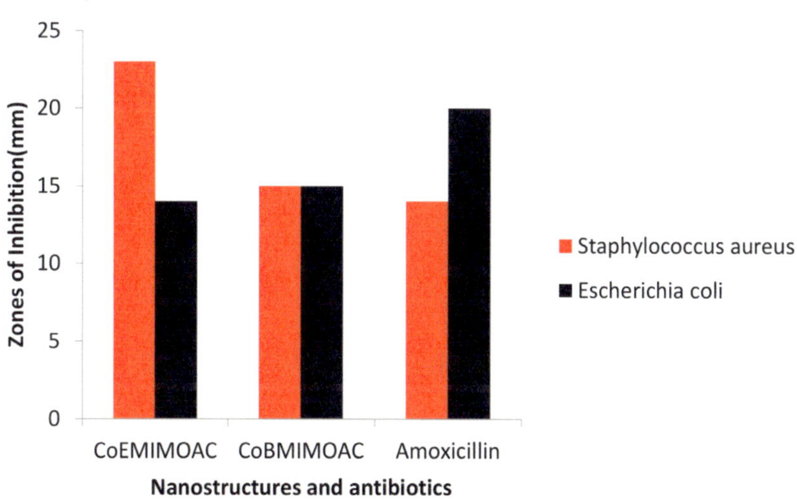

Figure 6.
Bar chart showing nanocrystals inhibition zone diameter for Staphylococcus aureus and Escherichia coli. Source: John et al. [43].

3. Conclusion

In this short review, the different protocols available for the synthesis of nanostructured layered double hydroxides and metal phosphonates formulated as UCIM-n have been discussed with their applications in the removal of scale forming ions from aqueous solutions. Reports of the work done in our research laboratory on green chemistry involving the use of plant extracts and ionic liquids as solvent and stabilizing agents in the syntheses of metal nanoparticles along with their antimicrobial activities have been presented.

Acknowledgements

This work was supported by The World Academy of Sciences for the Advancement of Science in developing countries (TWAS) under Research Grant No. 12-169 RG/CHE/AF/AC-G UNESCO FR: 3240271320 for which grateful acknowledgment is made. AAA is also grateful to the Royal Society of Chemistry for personal research grant. The assistance of Professor J-G. Mao of Fujian Institute of Research on the Structure of Matter, Chinese Academy of Sciences, in the XRD and SEM analyses is gratefully acknowledged.

Author details

Ayi A. Ayi*, Providence B. Ashishie, Emmanuel E. Khansi, Joseph O. Ogar, Chinyere A. Anyama and Bassey E. Inah
Inorganic Materials Research Laboratory, Department of Pure and Applied Chemistry, University of Calabar, Calabar, Nigeria

*Address all correspondence to: ayiayi72@gmail.com

References

[1] Caruso F, editor. Colloids and Colloid Assemblies: Synthesis, Modification, Organization and Utilization of Colloid Particles. Weinheim: Wiley VCH; 2004

[2] Goldenberg S. Dispersant 'May Make Deep-Water Horizon Oil Spill more Toxic'. The Guardian Newspaper. 5 May 2010

[3] Zahir F, Rizwi SJ, Haq SK, Khan RH. Low dose mercury toxicity and human health. Environmental Toxicology and Pharmacology. 2005;**20**:351-360

[4] USEP Agency. Nanotechnology for site remediation fact sheet. 2009

[5] Moffat G, Williams RA, Webb C, Stirling R. Selective separations in environmental and industrial processes using magnetic carrier technology. Minerals Engineering. 1994;7(8):1039-1056

[6] Ashishie PB, Inah BE, Ayi AA. Evaluation of antimicrobial activity of ionic liquid assisted synthesis of monometallic silver and bimetallic copper-silver nanoparticles. International Journal of Sciences. 2018;**7**(5):25-31

[7] Khansi EE, Anyama CA, Ayi AA, Onwuka JC. Use of nanostructured layered double hydroxides as nano filters in the removal of Fe^{2+} and Ca^{2+} ions from oil wells. Advances in Materials Science and Engineering. 2018;**2018**:1-7. DOI: 10.1155/2018/5306825

[8] Cheng H, Ma J, Zhao Z, Qi L. Hydrothermal preparation of uniform nanosize rutile and anatase particles. Chemistry of Materials. 1995;7:663-671

[9] Ge S, Shi X, Sun K, Li C, Uher C, Baker JR, et al. Facile hydrothermal synthesis of iron oxide nanoparticles with tunable magnetic properties. The Journal of Physical Chemistry. 2009;**C113**:13593-13599

[10] Kitamura Y, Okinaka N, Shibayama T, Mahaney OOP, Kusano D, Ohtani B, et al. Combustion synthesis of TiO_2 nanoparticles as photocatalyst. Powder Technology. 2007;**176**:93-98

[11] Jones AC, Chalker PR. Some recent developments in the chemical vapour deposition of electroceramics ocides. Journal of Physics D: Applied Physics. 2003;**36**:R80-R95

[12] Wang W, Lenggoro IW, Terashi Y, Kim TO, Okuyama K. One step synthesis of titanium oxide nanoparticles by spray pyrolysis of organic precursors. Materials Science and Engineering B. 2005;**123**:194-202

[13] Watson S, Beydoun D, Scott J, Amal R. Preparation of nanosized cryrstalline TiO_2 particles at low temperature for photocatalysis. Journal of Nanoparticle Research. 2004;**6**:193-207

[14] Douglas G, Shackleton M, Woods P. Hydrotalcite formation facilitates effective contaminant and radionuclide removal from acidic uranium mine barren lixiviant. Applied Geochemistry. 2014;**42**:27-37

[15] Douglas G. Contaminant removal from acidic mine pit water via in situ hydrotalcite formation. Applied Geochemistry. 2014;**51**:15-22

[16] Balsamo N, Mendieta S, Olivab M, Eimer G, Crivello M. Synthesis and characterization of metal mixed oxides from layered double hydroxides. Procedia Materials Science. 2012;**1**:506-513

[17] Cavani F, Trifiròa F, Vaccaria A. Hydrotalcite-type anionic clays: Preparation, properties and applications. Catalysis Today. 1991;**11**:173-301

[18] Rives V. Layered Double Hydroxides: Present and Future. New York, NY, USA: Nova Science Publishers; 2001

[19] Khan AI, O'Hare D. Intercalation chemistry of layered double hydroxides: Recent developments and applications. Journal of Materials Chemistry. 2002;**12**:3191-3198

[20] Sideris PJ, Nielsen UG, Gan Z, Grey CP. Mg/Al ordering in layered double hydroxides revealed by multinuclear NMR spectroscopy. Science. 2008;**321**:113-117

[21] Scarpellini D, Leonardi C, Mattoccia A, Di Giamberardino L, Medaglia PG, Mantini G, et al. Solution-grown Zn/Al layered double hydroxide nanoplatelets onto Al thin films: Fine control of position and lateral thickness. Journal of Nanomaterials. 2015;**2015**:1-8

[22] Yang M, Tuckley E, Buffet JC, O'Hare D. Rapid efficient phase pure synthesis of Ca_2AlNO_3 layered double hydroxide. Journal of Materials Chemistry. 2016;**A4**:500-504

[23] Everaert M, Warrinnier R, Baken S, Gustafsson JP, De Vos D, Smolders E. Phosphate-exchanged Mg–Al layered double hydroxides: A new slow release phosphate fertilizer. ACS Sustainable Chemistry & Engineering. 2016;**4**(8):4280-4287

[24] Bao W, Wang J, Wang Q, O'Hare D, Wan Y. Layered double hydroxide nano transporter for molecule delivery to intact plant cells. Scientific Reports. 2016;**6**:26738

[25] Kovačević D, Džakula BN, Hasenay D, Nemet I, Rončević S, Imre Dékány I, et al. Adsorption of arsenic on MgAl layered double hydroxide. Croatica Chemica Acta. 2013;**86**(3):273-279. DOI: 10.5562/cca2283

[26] Rives V, Ulibarri MA. Layered double hydroxides (LDH) intercalated with metal coordination compounds and oxometalates. Coordination Chemistry Reviews. 1999;**181**(1):61

[27] Kopka H, Beneke K, Lagaly G. Anionic surfactants between double metal hydroxide layers. Journal of Colloid and Interface Science. 1988;**123**:427-436

[28] Hornok V, Erdőhelyi A, Dékány I. Preparation ultrathin membranes by layer-by-layer deposition of layered double hydroxide (LDH) and polystyrene sulfonate (PSS). Colloid & Polymer Science. 2005;**283**:1050-1055

[29] Aradi T, Hornok V, Dékány I. Layered double hydroxide for ultrathin hybrid film preparation using layer-by-layer and spin coating methods. Colloids and Surfaces, A: Physicochemical and Engineering Aspects. 2007;**319**:116-121

[30] Tsyganok AI, Tsunoda T, Hamakawa S, Suzuki K, Takehira K, Hayakawa T. Dry reforming of methane over catalysts derived from nickel-containing Mg-Al layered double hydroxides. Journal of Catalysis. 2003;**213**:191-203

[31] Choy J-H, Choi S-J, Oh JM, Park T. Clay minerals and layered double hydroxides for novel biological applications. Applied Clay Science. 2007;**36**:122-132

[32] Bhattacharjee S, Dines TJ, James AA. Synthesis and application of layered double hydroxide hosted catalysts for stereoselective epoxidation using molecular oxygen or air. Journal of Catalysis. 2004;**225**:398-407

[33] Aisawa S, Ohnuma Y, Hirose K, Takahashi S, Hirahara H, Narita E. Structural characterization of arsenate ion exchanged MgAl-layered double hydroxide. Applied Clay Science. 2005;**28**:137-145

[34] Yang QZ, Yang J, Zhang CK. Synthesis and properties of cordycepin intercalates of Mg-Al-nitrate layered double hydroxides. International Journal of Pharmaceutics. 2006;**326**:148-152

[35] Zhang P, Shen D, Fan C, Kan AT, Tomson MB. Surfactant-assisted synthesis of metal-phosphonate inhibitor nanoparticles and its transport in porous media. In: A Conference Paper presented at the 2009 SPE International Oilfield Scale Conference in Woodland; SPE 1211552; 20-22 April 2009; Texas; 2009. pp. 1-12

[36] Shen D, Zhang P, Kan AT, Fu G, Farrell J, Tomson MB. Control placement of scale inhibitors in the formation with stable Ca-DTPMP nanoparticles suspension and its transport porous media. In: A Conference Paper presented at the 2008 SPE International Oilfield Scale Conference in Aberdeen, UK; 28-29 May; SPE 114063; 2008

[37] Ogar JO. Surfactant assisted synthesis of metal phosphonate nanoparticles: Structural characterization and application as oil field scale inhibitors [MSc thesis]. Nigeria: University of Calabar; 2015

[38] Bansal V, Bhargava SK. Ionic liquids as designer solvents for the synthesis of metal nanoparticles. In: Kokorin A, editor. Ionic Liquids: Theory, Properties, New Approaches. Croatia: InTech; 2011

[39] Khare V, Zhonghao Li Z, Alexandre Mantion A, Ayi AA, Sonkaria S, Voelk A, et al. Strong anion effects on gold nanoparticle formation in ionic liquids. Journal of Materials Chemistry. 2010;**20**:1332-1339

[40] Ayi AA, Khare V, Strauch P, Girard J, Fromm KM, Taubert A. On the chemical synthesis of titanium nanoparticles from ionic liquids. Monatshefte fuer Chemie. 2010;**141**:1273-1278. DOI: 10.1007/s00706-010-0403-4

[41] Ayi AA, Anyama CA, Khare V. On the synthesis of molybdenum nanoparticles under reducing conditions in ionic liquids. Journal of Materials. 2015;**2015**:1-7. DOI: 10.1155/2015/372716

[42] Ashishie PB, Anyama CA, Ayi AA, Oseghale CO, Adesuji ET, Labulo AH. Green synthesis of silver mono metallic and copper-silver bimetallic nanoparticles using *Kigelia africana* fruit extract and evaluation of their antimicrobial activities. International Journal of Physical Sciences. 2018;**13**(3):24-32. DOI: 10.5897/IJPS2017.4689

[43] John WE, Ayi AA, Anyama CA, Ashishie PB, Inah BE. On the use of methylimidazolium acetate ionic liquids as solvent and stabilizer in the synthesis of cobalt nanoparticles by chemical reduction method. Advanced Journal of Chemistry-Section A. 2019;**2**(2):175-183

Chapter 3

Two-Dimensional Nanomaterials

Zahra Rafiei-Sarmazdeh, Seyed Morteza Zahedi-Dizaji and Aniseh Kafi Kang

Abstract

Two-dimensional (2D) nanomaterials are composed of thin layers that may have a thickness of at least one atomic layer. Contrary to bulk materials, these nanomaterials have a high aspect ratio (surface-area-to-volume ratio) and therefore have many atoms on their surface. These atoms have a different function than internal atoms, and so the increase in the number of surface atoms leads to a change in the behavior of 2D nanomaterials. Graphene, as one of the most widely used and most important 2D materials, has unique properties that result in its widespread use in various industries. After successful performance of graphene in many applications and industry, it is expected that other two-dimensional materials will also have this capability. However, the use of other two-dimensional materials requires more time and effort.

Keywords: two-dimensional, nanomaterial, graphene, hexagonal boron nitride, chalcogenide

1. Introduction

Dimensional classification is one of the methods for classifying nanomaterials: the same chemical compounds can exhibit extraordinary different properties when they are configured in a zero (0D)-, one (1D)-, two (2D)-, and three (3D)-dimensional crystal structure [1]. In spite of the fact that there have been plenty of scientific reports on 0D [2], 1D [3–5] and, of course, 3D [6, 7], however, a limited number of researches on 2D nanomaterials are published.

2D nanomaterials are considered to be the thinnest nanomaterials due to their thickness and dimensions on macroscale/nanoscale. These nanomaterials have a layered structure with strong in-plane bonds and weak van der Waals (vdW) between layers. These ultrathin nanomaterials can be produced from laminated precursors described in the following sections. Although the ideal state is a single layer, but often these nanosheets are composed of few layers (less than ten layers). In recent years, 2D nanomaterials such as graphene, hexagonal boron nitride (hBN), and metal dichalcogenides (MX_2) have attracted a lot of attention due to their satisfactory properties and widespread uses in the electronics, optoelectronics, catalysts, energy storage facilities, sensors, solar cells, lithium batteries, composites, etc.

The schematic structure of graphene, boron nitride nanosheets, and tungsten diselenide (WSe_2) as a dichalcogenide has been illustrated in **Figure 1**. As shown, these compounds are configured in honeycomb structure, but the arrangement of the neighboring atoms in the upper and lower layers of 2D nanomaterials is

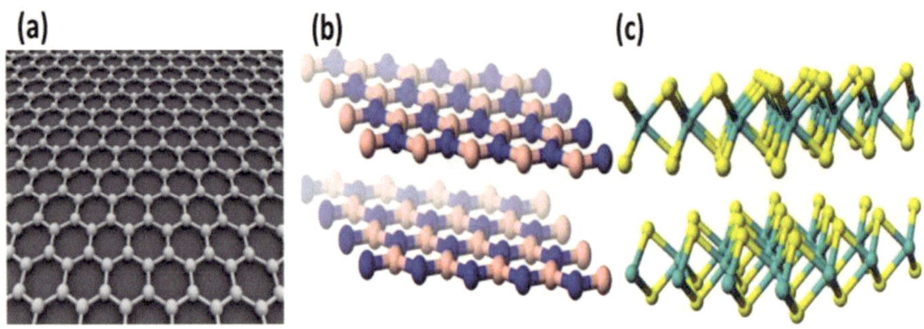

Figure 1.
The structure of (a) single layer of graphene with a lattice of carbon atoms, (b) boron nitride nanosheets with B in blue and N in pink, and (c) tungsten diselenide (WSe_2) with W in blue and Se in yellow [8].

different. In graphene, each carbon atom is next to another carbon atom in its upper and lower layers, while in the structure of BNNSs, each atom is located in the center of the benzene ring on the upper and lower layers. In the structure of dichalcogenides, each atomic layer of metal is sandwiched between two atomic layers of X.

In this chapter, the recent developments in the synthesis, properties of 2D nanomaterials especially graphene, and boron nitride nanosheets (BNNSs) are discussed. A comprehensive understanding of the properties and physics of these materials can be very effective in finding their application in the industry that is discussed in this chapter. The reported virtues and novelties of these nanomaterials are highlighted, and the current problems in their developing process are clarified.

2. Introduction of graphene

There are some well-known nanosheet materials with strong bonding on surface and poor bonding between layers such as graphene, BNNSs, and MX_2. Due to their specific structures, researchers have made great efforts to produce 2D nanosheets by exfoliating these layered compounds into distinct layers. A transmission electron microscopy image of crumpled monolayer of graphene is shown in **Figure 2**.

Graphene is the most famous of 2D nanosheets that is composed of carbon atoms in a hexagonal (honeycomb) configuration with sp^2-hybridized atoms

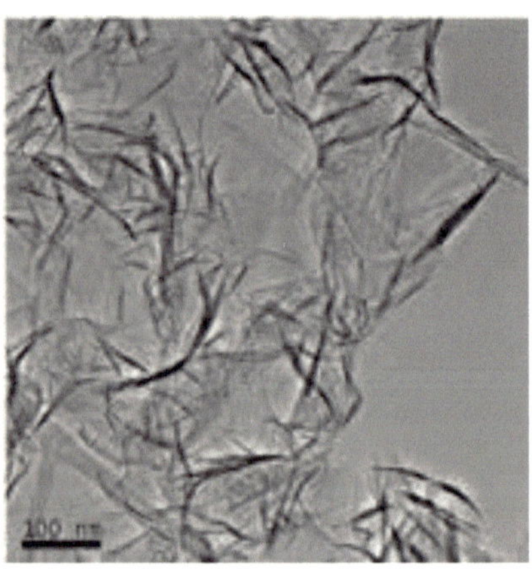

Figure 2.
The transmission electron microscope image of a crumpled graphene [9].

[10, 11]. Graphene is also the most important member of the multidimensional carbon material family that is formed by putting together carbon atoms and which included fullerene as the zero-dimensional nanomaterial (0D), carbon nanotubes as a one-dimensional nanomaterial (1D), and graphite as a three-dimensional nanomaterial (3D) (**Figure 3**) [12].

Graphene has a very weak absorption coefficient of 2.3% of white light, and so it's seen as a white powder. The surface area of graphene is 2630 m^2/gr, which is twice as much as carbon nanotubes with a surface area of 1315 m^2/gr. The legendary discovery of this compound in 2004 attracted a lot of attention and led to the discovery of great electronic properties, electron transfer capabilities, unprecedented impermeability, and high mechanical strength, excellent thermal, and electrical conduction.

Single-layer graphene is a substructure for the construction of carbon structures, which if placed on each other, produce 3D graphite. The attractive force between layers is van der Waals (vdW) force with a gap of 0.335 nm [14]. If single-layer graphene is rotated around the axis of the tube, the 1D carbon nanotube, and if they are wrapped spherically, it forms the 0D fullerene.

2.1 The synthesis of graphene

The numerous chemical and physical methods have been proposed for the production of different types of graphene (from single layer to few layer) based on top-down and bottom-up approaches. Chemical vapor deposition (CVD) and epitaxial growth [15, 16], plasma-enhanced chemical vapor deposition (PECVD) [15], mechanical cleavage [14, 17], Scotch® tape technique [17], chemical synthesis [18], liquid exfoliation [19, 20], etc. have been widely used to produce graphene.

2.2 Graphene properties

Considering the attention of scientists to graphene and the hope for its various applications in the near future, many research efforts have been devoted to understanding the structure and properties of graphene. Graphene is expected to

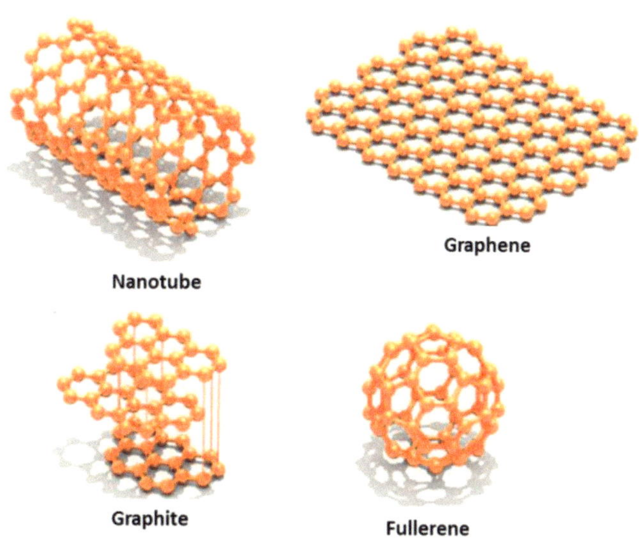

Figure 3.
Different forms of graphitic carbon [13].

consist of only single layer, but there is a significant attractive force to bind layers and to form two-layer or few-layer graphene. Two- and few-layer graphene consist of two and three to ten layers of these 2D nanosheets, respectively. The graphene structure, which contains more than ten of these 2D sheets, is considered to be "thick graphene" and is less of a concern for scientists. The status of graphene has changed from an unknown to a superstar in various fields of science and technology [21]. This is due to graphene's exceptional characteristics including high current density, ballistic transport, chemical inertness, high thermal conductivity, optical transmittance, and superficial hydrophobicity on a nanometer scale [14].

Single-layer graphene, as previously discussed in this chapter, is defined as a 2D nanosheet of carbon atoms that are arranged in a hexagonal network. Each sp^2-hybridized carbon atom is bonded to three another atoms with strong covalent bond (σ) that are configured in the hexagonal structure and also has a π orbital perpendicular to the sheet that forms π bond out of plane. These bonds can control the interaction between different layers of graphene in few-layer graphene [21].

Graphene is a semimetal or a semiconductor with a bandgap of zero and also has very high electron mobility at room temperature. Single-layer graphene has an unexpected high degree of transparency so that it absorbs $\pi\alpha \approx 2.3\%$ the incident white light, in which α is a substructure factor [7]. Single-layer graphene is also considered as one of the strongest materials. Given these mechanical properties, more applications in nanocomposite and coating industries are expected to be opened [21].

Graphene nanosheets are demonstrated to exhibit high transparency in UV-Vis and IR radiation and could be used to produce transparent electrode in solar cells [22]. Graphene has a good ability to functionalize with different functional groups in the form of covalent and noncovalent which leads to its solubility in different solvents. On the other hand, the high surface area of graphene provides a lot of area for loading of functional groups, which leads to reach a higher-level loading of targeting group in the surface, so graphene is considered as a suitable agent for drug delivery. In addition, the high surface area of graphene allow for development of targeted drug delivery systems [23].

2.3 Graphene applications

Different types of graphene, single-layer and few-layer, have potential applications in various fields. As stated above, graphene is the hardest and thinnest substance ever produced by human beings. Despite the fact that it has a dense structure, due to its very thin thickness, which is equal to the thickness of a carbon atom, it allows light to pass through and is highly transparent; it is also conductive, even more conductive than copper. Its ability to pass through heat and electricity makes it a new option for using on optical screens and computers.

It is 200–300 times stronger than steel and is even harder than diamond; however, it is very light and flexible. In addition, one of its properties is the great ability to move charge carriers. Electrons move relatively freely throughout graphene. With these features, graphene could be called supermassive, and it is expected that this material will create a revolution in the electronic, transistor, composite, coating, and sensor industries. Some examples of graphene applications can be:

- As reinforcement in composites instead of carbon fiber, this results in the creation of lighter and stronger aerocrafts and satellites.

- Used instead of semiconductor silicones in transistors due to superb conductivity properties. In this case, electrons can move 100 times faster than the electrons present in silicon, which is why potentially graphene have many applications in the electronics industry. This material is currently the main competitor of silicon [24].

- Embedding graphene in plastics to enable them to conduct electricity [25, 26].
- To increase the durability of batteries using graphene dust [27].
- Used in optical electronics [28].
- Make harder, stronger, and lighter plastics [29].
- As conductive transparent coating for solar cells and screens [30].
- Producing stronger implants (medical) [31].
- Create supercapacitors [32].
- Application in flexible touch screens and displays [33].
- Application in liquid crystal display (LCD) [34].
- Applications in light-emitting diodes (LEDs) and organic light-emitting diodes (OLEDs) [35].
- Making conductive inks for coating [36].

3. Introduction of hexagonal boron nitride

hBN is structurally similar to graphite and has hardness comparable to graphite. Since hBN is the isoelectric analog of graphite structure and shares very similar structural characteristics and many physical properties, is so-called white graphite. It is not present in nature and is synthesized.

Due to its unique properties, including high resistance to oxidation, high thermal conductivity, good thermal insulation, chemical inertness, excellent lubrication, non-toxicity, and environmental friendliness, hBN has diverse industrial applications in surface coatings, composites, lubricants, and insulators. Due to the impressive properties of nanoscale materials and the development of the application of nanomaterials in the industry, ongoing research is carried to develop new methods for synthesis of nanomaterials. However, until now, there is no ensured large-scale and high yield method to achieve a significant amount of boron nitride nanosheets (BNNSs).

Although researches on 2D nanomaterials have been began several decades ago, the wave of interest and attention to these materials get started in 2004 when Novoselov discovered single-layer graphene with superb electronic properties [1]. Many efforts have been made to achieve 2D materials including graphene, boron nitride, and several dichalcogenides. Boron nitride (BN) is one of the most promising systems ever to be the lightest compound of the three and four groups in the periodic table. BN is composed of equal numbers of N and B atoms, which are configured in hexagonal arrange, similar to carbon atoms in graphene. For naming, the term "single-layer BN" is used for monolayer of BN, and in the case of multilayers, is called BNNSs.

As shown in **Figure 4**, single-layer BN has honeycomb structure consisting of iso-electric borazine rings and benzene structures. B—N bonds have a covalent nature, but due to the electronegativity difference, these bonds have ionic properties with a length of 1.45 Å. The distance between the two centers of the borazine rings is 2.54 Å (compared with 2.46 Å for graphene). The edges of the plates could be zigzagged (boron or nitrogen on the edges) or armchair (nitrogen-boron on the edge) [37].

The lateral dimension of BNNSs is in range from several hundred nanometers to several 10 micrometers [39, 40]. The dimensions of the nanosheets are different depending on the synthesis method. **Figure 5** shows a transmission electron microscopy (TEM) micrograph possessing boron nitride plates, with a lateral dimension of 600 nm.

The single layers of BN can be placed on each other to form few-layer BNNSs. The vdW interaction between layers holds BN layers together, so that the distance between these sheets is 0.333 nm, while the layer's distance in carbon structures is about 0.337 nm [37].

The inorganic analog of graphene, sometimes assigned white graphene, is isoelectronic similar to graphene. However, due to electronegativity differences between the boron and the nitrogen atoms, π electrons are shifted into nitrogen atomic centers, forming the insulating materials [42, 43].

The arrangement of atomic layers in BN and its nanosheets differs with graphite and graphene. The arrangement in the graphene is called AB stacking mode, so that each carbon atom is located at the top of the center of the neighboring layer benzene ring. While in the layers of BN, the stacking mode is AA, and each atom at the upper and lower layers has nitrogen atoms due to polar-polar or electrostatic interactions [6]. Although the AA stacking mode is always observed in nanosheets obtained from the top-down approaches, this order is not always seen in bottom-up synthetic techniques [39, 40]. In addition, the calculations show that the B–N layers have relative displacement from AA to AB stacking mode, along the favorable energy [44].

Duo to the difference in electronegativity, B—N bonds have ionic characteristic which is compared to covalent C—C bonds in graphene. This can lead to lip-lip interactions between the layers, i.e., chemical interactions as bridges or spot-welds. This phenomenon helps to reduce energy and then stabilize the formation

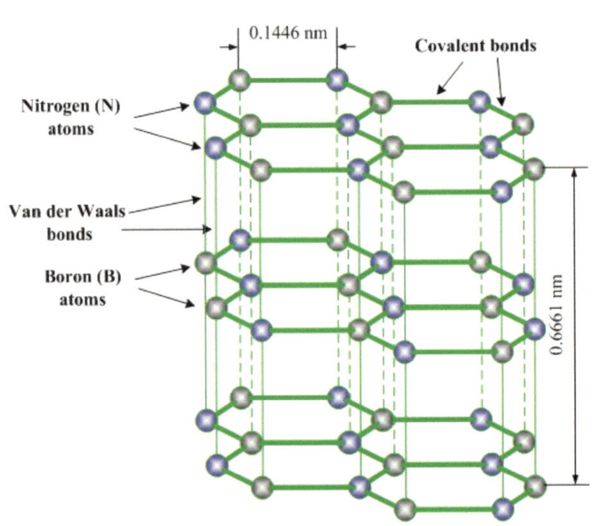

Figure 4.
Structural view of 2D BNNSs [38].

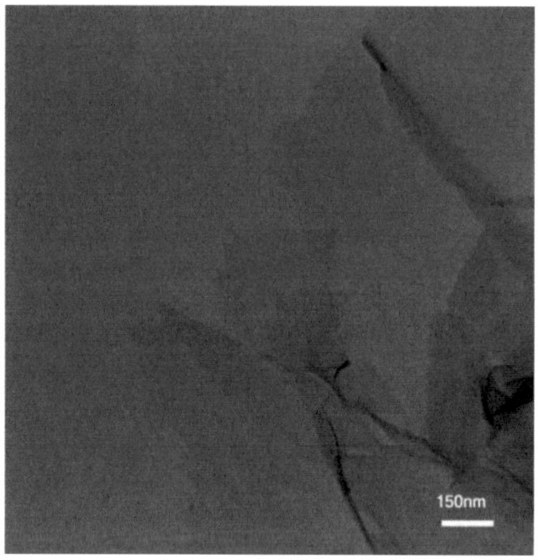

Figure 5.
The transmission electron microscope image of two relatively large BNNSs, which have been overlapped in the middle [41].

of few-layered BNNSs by reducing the number of dangling bonds at the edges, as well as reduction of frustration effect (forming B—B and N—N bonds instead of favorable B—N bonds) [45, 46]. Interestingly, such a strong interaction has only a negligible effect on the distance between the BN layers compared to graphene [6].

3.1 The preparation of BNNSs

2D nanosheets can be synthesized with two bottom-up and top-down approaches that relate to the synthesis of sheets from boron and nitrogen precursor [47, 48] and also the separation of layers. Chemical reaction [49] and CVD [39, 40, 50] are based on the former and micromechanical cleavage [51, 52], high-energy electron beam [53, 54], ball milling [55], and chemical [49, 56, 57]/liquid exfoliation [58–61] are based on the later approach.

Each technique has unique advantages for a specific application; however, there are always disadvantages in any way. In synthesis processes, a great amount of effort was put into preventing the formation of a strong chemical bond between the substrate and nanosheets. The crystallization process time, the nucleation on the substrate and the low density of critical nuclei are the important factors in synthetic methods. On the other hand, in the top-down view, exfoliation of layers is used. Nanosheets obtained from exfoliation usually have a higher crystallinity, but their lateral dimensions are limited by the material used. Also, due to lip-lip interactions between sheets, exfoliation of layers to isolate them is difficult. Therefore, the production of single layer is associated with a lot of problems. But in the bottom-up approach, there is a lot of control over the supply of thin nanosheets with high lateral dimensions. However, the crystallinity of obtained nanosheets is less than the exfoliation process.

3.2 The properties and applications of BNNSs

hBN has attracted many attentions due to its low density, high thermal conductivity, electrical insulation, high resistance to oxidation, low chemical efficiency, and low refractive index. BNNSs also inherit these properties, and in addition, they also have special properties due to high surface area.

3.2.1 Optical properties of BNNSs

BNNSs have not any absorption in the visible region but have absorption spectroscopy in the ultraviolet region [62]. Its commercial powders are white and its single crystal is transparent. Thin films obtained from chemical exfoliation or CVD also have high transparency [39, 40]. BNNS dispersion is often transparent at low concentrations and shows the Tyndall effect (the path of visible light inside the dispersion with laser light); at higher concentrations, the laser light is diffracted; and the dispersion is seen milky because of the lateral dimensions of nanosheets that are larger than the wavelength of laser light. Due to diffraction, there is any peak in the visible area. The measured extinction coefficient for absorption and diffraction of nanosheets is much smaller than that of graphene. The smaller lateral dimensions and more defects often influence on the optical bandgap, which is attributed to the absorption of the small distribution bandgap at the fermi surfaces and is produced due to the presence of defects [57]. However, hBN has two peaks in the 4 and 5.7 eVs: the first one related to the bandgap energy of BN, as a direct-gap semiconductor, and the latter is related to impurities and vacancy defects [63]. The lateral dimensions and number of layers affect the bandgap energy of nanosheets. For example, Rafiei-Sarmazdeh et al. [58] reported the absorption spectra of as-obtained BNNSs (2-nm-thick layer) at 204 nm (6.08 eV) that are related to the intrinsic excitement of BN structure and which is consistent with the reported results in other previous literatures [64, 65] and is also close to the bandgap energy predicted by theoretical calculus (6.0 eV) [66].

3.2.2 Thermal conductivity

The thermal conductivity for BNNSs is in range 300–2000 W/mK, which is comparable to graphene (1500–2500 W/mK). The difference in the conductivity may be due to the soft phonon modes of carbon sheets and the mass difference between boron and nitrogen [67]. Single BN layers have higher conductivity than multilayers, as the number of layer decreases and the phonon diffraction between layers reduces. As the number of layers increases, the conductivity decreases and converges to the conductivity of hBN. Although hBN has high conductivity and thermal capacity, recent studies have shown that its strong phonon diffraction leads to lower thermal conductivity than graphite. Therefore, the reduction of diffraction in BNNS leads to a significant increase in the conductivity (at room temperature $>$ 600 W/mK) [68].

3.2.3 Mechanical properties

The hardness for BNNS and graphene is 267 and 335 TPa, respectively [69]. Hence, BNNSs can be used as reinforcement for polymer composites. It has been shown that modulus and tensile strength for nanosheets (thickness of 1–2 nm) are in the range of 220–510 and 8–16 TPa [39]. For multilayer, it is expected that the main values are somewhat less than these values.

3.2.4 Lubricant properties

Another interesting case with BN materials is frictional properties. hBN and graphite are used as lubricants for many years. The lubricating properties result from the application of the external shear force on the weak forces between the layers and sliding. At the level of atomic layers, friction force microscopy (FFM) studies show that the friction properties of these nanosheets depend on their thickness.

Thin nanosheets show more friction due to increased out-of-plane deformation. More studies have shown that reversible dynamic wrinkling on the top surface of most layers is created by frictional force [70]. Then, BNNSs and graphene could be used as lubricating additive.

3.2.5 The neutron absorption

Boron is introduced as one of the most important neutron absorbers due to its high neutron absorption cross section. The compounds containing of boron are good neutron absorber. In the middle, hBN and, of course, BNNSs are better absorbers due to layer structure and larger surface area that is exposed to neutron beam than other BN structures (such as nanotube, nanoparticle, etc.). They are used in nuclear shielding [71] and boron neutron therapy [72].

4. Introduction of metal dichalcogenides

In recent years, metal dichalcogenides (MX_2) have attracted a lot of attention, like other 2D materials where M is transition metals and X_s are S, Se, and Te such as WSe_2, molybdenum disulfide (MoS_2), tellurium disulfide (TeS_2), etc. MX_2 can be a semiconductor or metal depending on the oxidation state of their metal atoms. As graphene-like compounds, they have similarities to graphene, resulting in new opportunities for detecting and building sensors, lithium batteries, optoelectronics, and energy storage. Many researchers have shown that the exfoliated layers of MX_2 have large energy of bandgap and are semiconductors. In addition, they have the properties of fluorescence and photoluminescence [73].

There are currently only a limited number of reports on single-layer and few-layer group IV–VI and III–VI layered compounds. Therefore, the 2D properties of these materials are largely unknown. However, MX_2 shows a variety of electrical and optical properties that are suitable for catalysts, nanotribology, optoelectronics, and lithium-ion batteries. These materials are expected to exhibit extraordinary properties after reaching a thickness to single layer or multilayer [74].

4.1 The synthesis of TMD

The synthesis routes to MX_2 are similar to other 2D nanomaterials that are based on two top-down and bottom-up approaches. Methods such as chemical synthesis, CVD, mechanical cleavage, and liquid exfoliation are performed to produce of these nanomaterials [75].

4.2 The application of TMD

These layered compounds, like other 2D nanomaterials, have interesting optical, electrical, photovoltaic, and catalytic properties. These compounds are considered as the next generation of flexible and ultrathin photoelectric devices. Also, MX_2 has a suitable photovoltaic response to laser excitation and also used as catalyst in the hydrogen evolution reactions (HER).

5. Future outlook

One of the major problems with 2D nanostructures is to achieve a large-scale synthesis method for producing high quality, large surface area, high crystallinity, and free from any impurity nanosheets. Different methods have been reported to synthesize these nanosheets so far, which generally suffer from problems such as impurities, low crystallinity, low lateral dimension, and little yield, which limit the use of these 2D nanosheets in the industry. At present, many studies focus on improving the synthesis methods of these nanosheets.

6. Conclusions

This chapter is an attempt to better realize 2D nanostructures, especially graphene and boron nitride, and getting to know the synthesis methods and the application of these materials in various fields. Recently, there have been advances in the production and application of 2D nanostructures, especially graphene and boron nitride.

In general, the properties and applications of nanostructures are determined by their structure and morphology. The large surface area, high aspect ratio, and much number of atoms on the surface provide the special properties to these 2D nanomaterials, such as thermal and electric conductivity, lubricating, mechanical characteristic, etc. These impressive properties allow them to be used in fields such as coatings, electrical and optoelectronic devices, composites, etc. However, the low efficiency of the synthesis methods, especially boron nitride, which has an ionic nature, creates some of the limitations that researchers are trying to overcome.

Conflict of interest

The authors declare that they have no conflict of interest.

Author details

Zahra Rafiei-Sarmazdeh[1*], Seyed Morteza Zahedi-Dizaji[1] and Aniseh Kafi Kang[2]

1 Plasma and Nuclear Fusion Research School, Nuclear Science and Technology Research Institute, Tehran, Iran

2 Department of Physics, Shahid Bahonar University of Kerman, Kerman, Iran

*Address all correspondence to: zrafiei@alumni.ut.ac.ir

References

[1] Novoselov KS, Jiang D, Schedin F, Booth TJ, Khotkevich VV, Morozov SV, et al. Two-dimensional atomic crystals. Proceedings of the National Academy of Sciences of the United States of America. 2005;**102**:10451-10453. DOI: 10.1073/pnas.0502848102

[2] Golberg D, Bando Y, Stéphan O, Kurashima K. Octahedral boron nitride fullerenes formed by electron beam irradiation. Applied Physics Letters. 1998;**73**:2441-2443. DOI: 10.1063/1.122475

[3] Li L, Chen Y, Stachurski ZH. Boron nitride nanotube reinforced polyurethane composites. Progress in Natural Science: Materials International. 2013;**23**:170-173. DOI: 10.1016/j.pnsc.2013.03.004

[4] Chopra NG, Luyken RJ, Cherrey K, Crespi VH, Cohen ML, Louie SG, et al. Boron nitride nanotubes. Science. 1995;**269**:966-967. DOI: 10.1126/science.269.5226.966

[5] André C, Guillaume YC. Boron nitride nanotubes and their functionalization via quinuclidine-3-thiol with gold nanoparticles for the development and enhancement of the HPLC performance of HPLC monolithic columns. Talanta. 2012;**93**:274-278. DOI: 10.1016/j.talanta.2012.02.033

[6] Hod O. Graphite and hexagonal boron-nitride have the same interlayer distance. Why? Journal of Chemical Theory and Computation. 2012;**8**:1360-1369. DOI: 10.1021/ct200880m

[7] Kuzmenko AB, Heumen EV, Carbone F, Marel D. Universal optical conductance of graphite. Physical Review Letters. 2008;**100**:117401-117404. DOI: 10.1103/physrevlett.100.117401

[8] 2D materials: Graphene, hBN and WSe2 [Internet]. 2016. Available from: https://www.cei.washington.edu

[9] Graphene Support Films for TEM [Internet]. 2019. Available from: https://www.emsdiasum.com/microscopy/products/graphene/graphene_tem.aspx

[10] Geim AK. Graphene: Status and prospects. Science. 2009;**324**:1530-1534. DOI: 10.1126/science.1158877

[11] Katsnelson MI. Graphene: Carbon in two dimensions. Materials Today. 2007;**10**:20-27. DOI: 10.1016/S1369-7021(06)71788-6

[12] Geim AK, Novoselov KS. The rise of graphene. Nature Materials. 2007;**6**:183-191. DOI: 10.1038/nmat1849

[13] Sinitskii A. Tour JM. Graphene Electronics, Unzipped [Internet]. 2010. Available from: https://spectrum.ieee.org/semiconductors/materials/graphene-electronics-unzipped

[14] Geim AK, Kim P. Carbon wonderland. Scientific American. 2008;**298**:90-97. DOI: 10.1038/scientificamerican0408-90

[15] Rao CNR, Sood AK, Subrahmanyam KS, Govindaraj A. Graphene: The new two-dimensional nanomaterial. Angewandte Chemie International Edition. 2009;**48**:7752-7777. DOI: 10.1002/anie.200901678

[16] Shao Y, Wang J, Wu H, Liu J, Aksay IA, Lin Y. Graphene based electrochemical sensors and biosensors: A review. Electroanalysis. 2010;**22**:1027-1036. DOI: 10.1002/elan.200900571

[17] Novoselov KS et al. Electric field effect in atomically thin carbon films. Science. 2004;**306**:666-669. DOI: 10.1126/science.1102896

[18] Park S, Ruoff RS. Chemical methods for the production of graphenes. Nature Nanotechnology.

2009;**4**:217-224. DOI: 10.1038/nnano.2009.58

[19] Sampath S, Basuray AN, Hartlieb KJ, Aytun T, Stupp SI, Stoddart JF. Direct exfoliation of graphite to graphene in aqueous media with diazaperopyrenium dications. Advanced Materials. 2013;**25**:2740-2745. DOI: 10.1002/adma.201205157

[20] Xia ZY et al. The exfoliation of graphene in liquids by electrochemical, chemical, and sonication-assisted techniques: A nanoscale study. Advanced Functional Materials. 2013;**23**:4684-4693. DOI: 10.1002/adfm.201203686

[21] Wonbong C, Indranil L, Raghunandan S, Soo KY. Synthesis of graphene and its applications: A review. Critical Reviews in Solid State and Materials Sciences. 2010;**35**:52-71. DOI: 10.1080/10408430903505036

[22] Peres NMR. The electronic properties of graphene and its bilayer. Vacuum. 2009;**83**:1248-1252. DOI: 10.1016/j.vacuum.2009.03.018

[23] Niyogi S, Bekyarova E, Itkis ME, McWilliams JL, Hamon MA, Haddon RC. Solution properties of graphite and graphene. Journal of the American Chemical Society. 2006;**128**:7720-7721. DOI: 10.1021/ja060680r

[24] Petrone N, Chari T, Meric I, Wang L, Shepard KL, Hone J. Flexible graphene field-effect transistors encapsulated in hexagonal boron nitride. ACS Nano. 2015;**9**:8953-8959. DOI: 10.1021/acsnano.5b02816

[25] Pang H, Chen T, Zhang G, Zeng B, Li ZM. An electrically conducting polymer/graphene composite with a very low percolation threshold. Materials Letters. 2010;**64**:2226-2229. DOI: 10.1016/j.matlet.2010.07.001

[26] Luong ND et al. Enhanced mechanical and electrical properties of polyimide film by graphene sheets via in situ polymerization. Polymer. 2011;**52**:5237-5242. DOI: 10.1016/j.polymer.2011.09.033

[27] Zhang X et al. Electrospun TiO_2–graphene composite nanofibers as a highly durable insertion anode for lithium ion batteries. The Journal of Physical Chemistry C. 2012;**116**:14780-14788. DOI: 10.1021/jp302574g

[28] Liu M et al. A graphene-based broadband optical modulator. Nature. 2011;**474**:64-67. DOI: 10.1038/nature10067

[29] Ning H et al. Interlaminar mechanical properties of carbon fiber reinforced plastic laminates modified with graphene oxide interleaf. Carbon. 2015;**91**:224-233. DOI: 10.1016/j.carbon.2015.04.054

[30] Li X et al. Transfer of large-area Graphene films for high-performance transparent conductive electrodes. Nano Letters. 2009;**9**:4359-4363. DOI: 10.1021/nl902623y

[31] Oyefusi A et al. Hydroxyapatite grafted carbon nanotubes and graphene nanosheets: Promising bone implant materials. Spectrochimica Acta Part A: Molecular and Biomolecular Spectroscopy. 2014;**132**:410-416. DOI: 10.1016/j.saa.2014.04.004

[32] Yoo JJ et al. Ultrathin planar graphene supercapacitors. Nano Letters. 2011;**11**:1423-1427. DOI: 10.1021/nl200225j

[33] Wang J, Liang M, Fang Y, Qiu T, Zhang J, Zhi L. Rod-coating: Towards large-area fabrication of uniform reduced graphene oxide films for flexible touch screens. Advanced Materials. 2012;**24**:2874-2878. DOI: 10.1002/adma.201200055

[34] Kim JE et al. Graphene oxide liquid crystals. Angewandte Chemie.

2011;**123**:3099-3103. DOI: 10.1002/anie.201004692

[35] Gunho J et al. Large-scale patterned multi-layer graphene films as transparent conducting electrodes for GaN light-emitting diodes. Nanotechnology. 2010;**21**:175201-175207. DOI: 10.1088/0957-4484/21/17/175201

[36] Karagiannidis PG et al. Microfluidization of graphite and formulation of graphene-based conductive inks. ACS Nano. 2017;**11**:2742-2755. DOI: 10.1021/acsnano.6b07735

[37] Lin Y, Connell JW. Advances in 2D boron nitride nanostructures: Nanosheets, nanoribbons, nanomeshes, and hybrids with graphene. Nanoscale. 2012;**4**:6908-6939. DOI: 10.1039/c2nr32201c

[38] Kopeliovich D. Boron nitride as solid lubricant [Internet]. 2012. Available from: http://www.substech.com/dokuwiki/doku.php?id=boron_nitride_as_solid_lubricant&s=hexagonal%20boron%20nitride

[39] Song L et al. Large scale growth and characterization of atomic hexagonal boron nitride layers. Nano Letters. 2010;**10**:3209-3215. DOI: 10.1021/nl1022139

[40] Shi Y et al. Synthesis of few-layer hexagonal boron nitride thin film by chemical vapor deposition. Nano Letters. 2010;**10**:4134-4139. DOI: 10.1021/nl1023707

[41] Rafiei-Sarmazdeh Z, Ahmadi SJ, Jafari SH, Kasesaz Y. Large-Scale production and functionalization of two-dimensional boron nitride nanosheets. In: Proceedings of the 6th International Conference on Nanostructures (ICNS6). Kish Island, Iran; 7-10 March 2016

[42] Golberg D et al. Boron nitride nanotubes and nanosheets. ACS Nano. 2010;**4**:2979-2993. DOI: 10.1021/nn1006495

[43] Golberg D, Bando Y, Tang C, Zni C. Boron nitride nanotubes. Advanced Materials. 2007;**19**:2413-2432. DOI: 10.1002/adma.200700179

[44] Marom N et al. Stacking and registry effects in layered materials: The case of hexagonal boron nitride. Physical Review Letters. 2010;**105**:046801-046804. DOI: 10.1103/PhysRevLett.105.046801

[45] Erickson KJ et al. Longitudinal splitting of boron nitride nanotubes for the facile synthesis of high quality boron nitride nanoribbons. Nano Letters. 2011;**11**:3221-3226. DOI: 10.1021/nl2014857

[46] Blase X, De Vita A, Charlier JC, Car R. Frustration effects and microscopic growth mechanisms for BN nanotubes. Physical Review Letters. 1998;**80**:1666-1669. DOI: 10.1103/PhysRevLett.80.1666

[47] Motojima S, Tamura Y, Sugiyama K. Low temperature deposition of hexagonal BN films by chemical vapour deposition. Thin Solid Films. 1982;**88**:269-274. DOI: 10.1016/0040-6090(82)90056-6

[48] Yu J et al. Vertically aligned boron nitride nanosheets: Chemical vapor synthesis, ultraviolet light emission, and superhydrophobicity. ACS Nano. 2010;**4**:414-422. DOI: 10.1021/nn901204c

[49] Nag A, Raidongia K, Hembram KPSS, Datta R, Waghmare UV, Rao CNR. Graphene analogues of BN: Novel synthesis and properties. ACS Nano. 2010;**4**:1539-1544. DOI: 10.1021/nn9018762

[50] Gao R et al. High-yield synthesis of boron nitride nanosheets with strong

ultraviolet cathodoluminescence emission. The Journal of Physical Chemistry C. 2009;**113**:15160-15165. DOI: 10.1021/jp904246j

[51] Pacilé D, Meyer JC, ÇÖ G, Zettl A. The two-dimensional phase of boron nitride: Few-atomic-layer sheets and suspended membranes. Applied Physics Letters. 2008;**92**:133107-133109. DOI: 10.1063/1.2903702

[52] Kubota Y, Watanabe K, Tsuda O, Taniguchi T. Hexagonal boron nitride single crystal growth at atmospheric pressure using Ni–Cr solvent. Chemistry of Materials. 2008;**20**:1661-1663. DOI: 10.1021/cm7028382

[53] Jin C, Lin F, Suenaga K, Iijima S. Fabrication of a freestanding boron nitride single layer and its defect assignments. Physical Review Letters. 2009;**102**:195505. DOI: 10.1103/PhysRevLett.102.195505

[54] Yu J, Huang X, Wu C, Wu X, Wang G, Jiang P. Interfacial modification of boron nitride nanoplatelets for epoxy composites with improved thermal properties. Polymer. 2012;**53**:471-480. DOI: 10.1016/j.polymer.2011.12.040

[55] Li LH, Chen Y, Behan G, Zhang H, Petravic M, Glushenkov AM. Large-scale mechanical peeling of boron nitride nanosheets by low-energy ball milling. Journal of Materials Chemistry. 2011;**21**:11862-11866. DOI: 10.1039/C1JM11192B

[56] Lin Y, Williams TV, Connell JW. Soluble, exfoliated hexagonal boron nitride nanosheets. Journal of Physical Chemistry Letters. 2010;**1**:277-283. DOI: 10.1021/jz9002108

[57] Lin Y, Williams TV, Cao W, Elsayed-Ali HE, Connell JW. Defect functionalization of hexagonal boron nitride nanosheets. Journal of Physical Chemistry C. 2010;**114**:17434-17439. DOI: 10.1021/jp105454w

[58] Rafiei-Sarmazdeh Z, Jafari SH, Ahmadi SJ, Zahedi-Dizaji SM. Large-scale exfoliation of hexagonal boron nitride with combined fast quenching and liquid exfoliation strategies. Journal of Materials Science. 2016;**51**:3162-3169. DOI: 10.1007/s10853-015-9626-4

[59] Nicolosi V, Chhowalla M, Kanatzidis MG, Strano MS, Coleman JN. Liquid exfoliation of layered materials. Science. 2013;**340**:1226419-1226436. DOI: 10.1126/science.1226419

[60] Kostecki M, Olszyna AR, Sokołowska A. Liquid exfoliation–new low-temperature method of nanotechnology. Materials Science. 2013;**31**:165-172. DOI: 10.2478/s13536-012-0086-0

[61] Hernandez Y et al. High-yield production of graphene by liquid-phase exfoliation of graphite. Nature Nanotechnology. 2008;**3**:563-568. DOI: 10.1038/nnano.2008.215

[62] Wang J, Ma F, Sun M. Graphene, hexagonal boron nitride, and their heterostructures: Properties and applications. RSC Advances. 2017;**7**:16801-16822. DOI: 10.1039/C7RA00260B

[63] Watanabe K, Taniguchi T, Kanda H. Direct-bandgap properties and evidence for ultraviolet lasing of hexagonal boron nitride single crystal. Nature Materials. 2004;**3**:404-409. DOI: 10.1038/nmat1134

[64] Coleman JN et al. Two-dimensional nanosheets produced by liquid exfoliation of layered materials. Science. 2011;**331**:568-571. DOI: 10.1126/science.1194975

[65] Lin Y, Williams TV, Xu TB, Cao WH, Elsayed-Ali E, Connell JW. Aqueous dispersions of few-layered and monolayered hexagonal boron nitride nanosheets from sonication-assisted hydrolysis: Critical role of water. Journal

of Physical Chemistry C. 2011;**115**: 2679-2685. DOI: 10.1021/jp110985w

[66] Blase X, Rubio A, Louie SG, Cohen ML. Quasiparticle band structure of bulk hexagonal boron nitride and related systems. Physical Review B. 1995;**51**:6868-6875. DOI: 10.1103/PhysRevB.51.6868

[67] Sevik C, Kinaci A, Haskins JB, ÇağIn T. Characterization of thermal transport in low-dimensional boron nitride nanostructures. Physical Review B: Condensed Matter and Materials Physics. 2011;**84**:085409-085415. DOI: 10.1103/PhysRevB.84.085409

[68] Bhattacharya A, Bhattacharya S, Das GP. Strain-induced band-gap deformation of H/F passivated graphene and h-BN sheet. Physical Review B: Condensed Matter and Materials Physics. 2011;**84**:075454-075466. DOI: 10.1103/PhysRevB.84.075454

[69] Şahin H et al. Monolayer honeycomb structures of group-IV elements and III-V binary compounds: First-principles calculations. Physical Review B: Condensed Matter and Materials Physics. 2009;**80**:155453-155464. DOI: 10.1103/PhysRevB.80.155453

[70] Barboza APM, Chacham H, Neves BRA. Universal response of single-wall carbon nanotubes to radial compression. Physical Review Letters. 2009;**102**:025501-025504. DOI: 10.1103/PhysRevLett.102.025501

[71] Harrison C, Weaver S, Bertelsen C, Burgett E, Hertel N, Grulke E. Polyethylene/boron nitride composites for space radiation shielding. Journal of Applied Polymer Science. 2008;**109**:2529-2538. DOI: 10.1002/app.27949

[72] Hawthorne MF, Lee MW. A critical assessment of boron target compounds for boron neutron capture therapy. Journal of Neuro-Oncology. 2003;**62**: 33-45. DOI: 10.1007/BF02699932

[73] Mak KF, Lee C, Hon J, Shan J, Heinz TF. Atomically thin MoS2: A new direct-gap semiconductor. Physical Review Letters. 2010;**105**:136805-136808. DOI: 10.1103/PhysRevLett.105.136805

[74] Xu M, Liang T, Shi M, Chen H. Graphene-like two-dimensional materials. Chemical Reviews. 2013;**113**:3766-3798. DOI: 10.1021/cr300263a

[75] Ruitao Lv, Robinson JA, Schaak RE, Sun D, Sun Y, Mallouk TE, et al. Transition metal dichalcogenides and beyond: Synthesis, properties, and applications of single- and few-layer nanosheets. Accounts of Chemical Research. 2015;**48**:56-64. DOI: 10.1021/ar5002846

Chapter 4

Low-Dimensional ZnO Nanostructures: Fabrication, Optical Properties, and Applications for Dye-Sensitized Solar Cells

Hsin-Ming Cheng and Shun-Wei Liu

Abstract

Zinc oxide nanostructure has a wide bandgap energy of 3.37 eV and a large exciton binding energy of 60 meV at room temperature. It is certainly a promising material for photonic devices in the ultraviolet to blue wavelength range. ZnO-related materials are also expected to construct the exciton as well as polariton lasers owing to their excitonic-stimulated emission and laser behavior under optically pumping can be obtained at ambient temperature. Because of the optical losses, including not only nonradiative recombination centers but also traps of excitons, the high quality of ZnO becomes even more imperative in the excitonic lasing processes. In the present chapter, ZnO nanowire structures via a low-pressure vapor-phase deposition and a simple solvothermal method will be presented. The one-dimensional ZnO nanowires could afford a direct conduction pathway to significantly enhance the overall efficiency of the dye-sensitized solar cells. Furthermore, this content will demonstrate how to employ the hierarchical structure of the ZnO nanoparticles, fabricated from sol-gel method, which could promote light scattering, thus, enhancing photon absorption and the overall solar conversion efficiency. The aim of this chapter is to present the correlation between the fundamental properties of ZnO nanostructures and their photovoltaics performances.

Keywords: zinc oxide, II–VI semiconductor, nanostructure, nanowire, nanoparticle, dye-sensitized solar cell

1. Introduction

Low-dimensional ZnO nanoparticles (NPs), quantum dots (QDs), and ZnO nanowires (NWs) have attracted much attention due to their good crystal quality, chemical stability, and unique optical properties. ZnO nanowire is also expected to play an important role because their manifold properties are interconnected as functional units in the fabrication of electronics, photonics, photocatalysts, and piezotronics with nanoscale dimensions. ZnO QDs and NPs are of great interest because of the three-dimensional confinement of carrier, and phonon leads not only continuous tuning of the optoelectronic properties but also improvement in device performance. As a wide-bandgap semiconductor, ZnO has been reported as an alternative for dye-sensitized solar cells (DSCs) because ZnO offers a large direct bandgap which is close to TiO_2 and even higher electron mobility (155 $cm^2 V^{-1} s^{-1}$) for the high-quality thin film [1]. ZnO provides a promising alternative for improving the performance of the photoelectrode in DSCs because ZnO can be tailored to various nanostructures. In the present proposal, first, I will present the ZnO nanowire structures via a low-pressure vapor-phase deposition and a simple solvothermal method. The one-dimensional ZnO NWs could simultaneously afford a direct conduction pathway to significantly enhance the overall efficiency of the DSCs [2, 3]. Next, I also will demonstrate how to employ the hierarchical structure of the ZnO NPs, fabricated from sol-gel method, which would promote light scattering through the presence of secondary colloidal spheres, thus, enhancing photon absorption to improve the short-circuit current density and the overall light conversion efficiency.

2. ZnO nanostructures from solvothermal method

The process of the growth of the ZnO NWs is similar to Law et al. [2]. Arrays of ZnO NWs were synthesized on fluorine-doped tin oxide (FTO) substrates that were first cleaned thoroughly by sonication with acetone/ethanol and then coated with a thin film of ZnO QDs, 3–4 nm in diameter, by dip-coating in a 0.005 M $Zn(OAc)_2$-concentrated ethanol solution. The arrays of ZnO NWs were synthesized on seeded FTO substrates by immersing the seeded substrates in aqueous solutions containing 0.06 M zinc nitrate hydrate, 0.06 M hexamethylenetetramine (HMTA), and 7.5 mM polyethylenimine (PEI) at 95°C for 2.5–7.5 hours. **Figure 1** shows the tilted-view SEM images of the ZnO NWs on FTO layers and alumina-doped ZnO layers (AZO, 50~200 Ω per square), respectively. It seems that the aspect ratio (length against diameter) of nanowires can be increased when using PEI. The length and diameter of the PEI-absent ZnO NWs can be controlled in the range of about 300–500 nm and 300–400 nm, respectively. But, the length and diameter of the PEI-present ZnO NWs can be adjusted in the range of about 500–700 nm and 150–250 nm, respectively. The lengths of ZnO NWs are also prolonged while increasing the growth times. For example, the length and diameter of the ZnO NWs were in the range of about 2–3 μm and 250–350 nm, respectively, while the growth time is of 7.5 hours. Furthermore, the ZnO nanostructures with different morphology can be observed while using various substrates. The density and diameter of ZnO NWs seem to be influenced while using AZO substrates. The length and diameter of the ZnO NWs were in the range about 0.2–1.2 μm and 200–500 nm, respectively. By sputtering technique, it is reasonable to presume that the grain size of thick AZO films (200–300 nm) is bigger than dip-coating method, which caused the increase in the NW diameter and the good orientation as well. However, the fluctuation of diameter for ZnO NWs is huge while using AZO substrates, which could be related to the uniformity of AZO under layers.

Beyond the ZnO NWs, the impressive branched ZnO NWs have been fabricated successfully from solvothermal method [4]. First, the arrays of ZnO NWs were synthesized on seeded FTO substrates by immersing the seeded substrates in aqueous solutions containing 0.08 M zinc nitrate hydrate, 0.08 M HTMA, and 12 mM PEI at 95°C for 10 hours. Second, the ZnO NW substrate obtained from the first step were re-coated with seed layers of ZnO NPs by dip-coating in a 0.005 M $Zn(OAc)_2$ in ethanol. Then the branched NWs were grown by immersing the seeded ZnO NWs for a 5-hour duration in an aqueous solution containing 0.02 M zinc nitrate hydrate, 0.02 M HMTA, and 3 mM PEI at 95°C ambient. The final products were immediately rinsed with deionized water and baked in air at 450°C for 30 minutes to

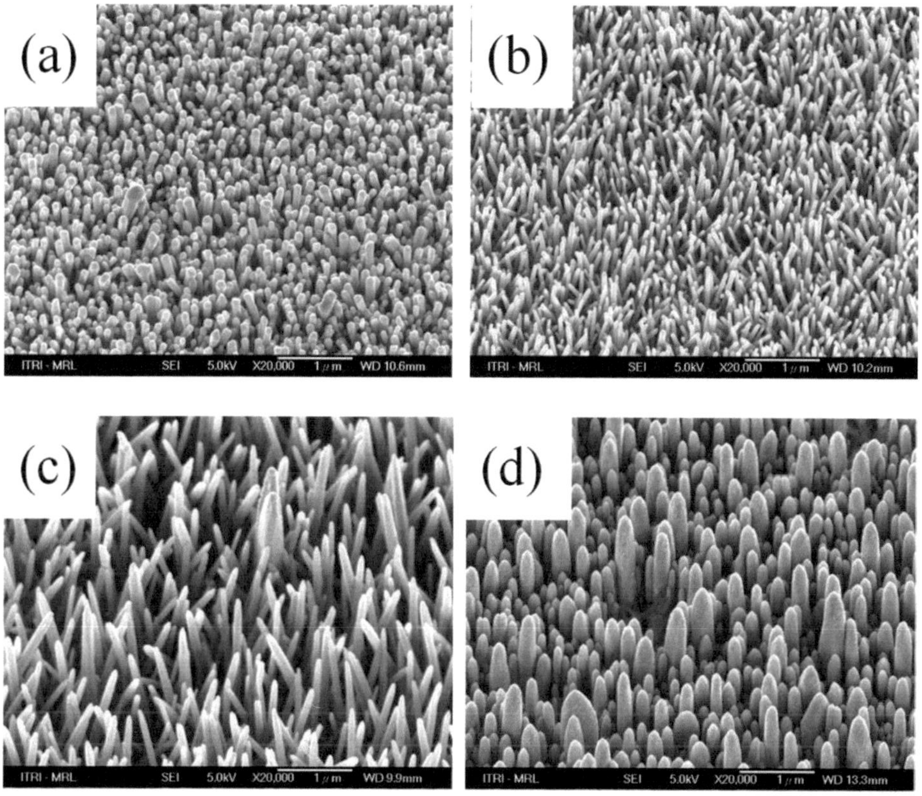

Figure 1.
The SEM images of the ZnO nanowires and nanorods with different growth conditions (a) FTO substrate, without PEI, grown 2.5 hours. (b) FTO substrate, with PEI, grown 2.5 hours. (c) FTO substrate, with PEI, grown 7.5 hours. (d) AZO substrate, with PEI and grown 7.5 hours.

remove any residual organics. The evolution of the ZnO NWs to the branched ZnO NWs is illustrated in **Figure 2a** and its corresponding FESEM images **Figure 2b–d**. By using a solvothermal method, the bare ZnO NWs with slight vertical off-alignment were grown perpendicularly on the FTO substrate, as shown in **Figure 2b**. Through the pre-coating process on the ZnO NWs, little ZnO crystallites with diameter 10–20 nm were formed on the backbone NWs, as shown in **Figure 2c**. After the second growth step, radial secondary ZnO branches emanated from the seeds, as shown in **Figure 2d**. The entire substrate in which the backbone nanowire has a density of about 7×10^8 cm^{-2} was covered with branched ZnO nanostructures. The backbone NWs have length and diameter in the range of 7–8 μm and 150–250 nm, respectively, whereas the secondary branches have length and diameter ranging from 100 to 300 nm and 20 to 50 nm, respectively. By the infiltration of moderately concentrated $Zn(OAc)_2$ solution into interstitial voids between backbone ZnO NWs, the branched ZnO NWs can successfully be fabricated.

Further structural characterizations of the branched ZnO NWs were performed by TEM. **Figure 3a** reveals the secondary ZnO branches which were grown on the side walls of nanowire backbone with different radial angles. The evidence confirmed the secondary ZnO branches were not the derivatives of ZnO NWs but definitely originated from the small crystallite ZnO seeds via the pre-coating process. Different from the other reports of comblike ZnO nanostructures which demonstrated the monolithically single-crystalline relationship between the branches and backbone nanowires [5, 6], in the present mechanism, the secondary ZnO branches were derived on the ZnO seeds in spite of the coordinate crystal relationship. The magnified intersection area of ZnO

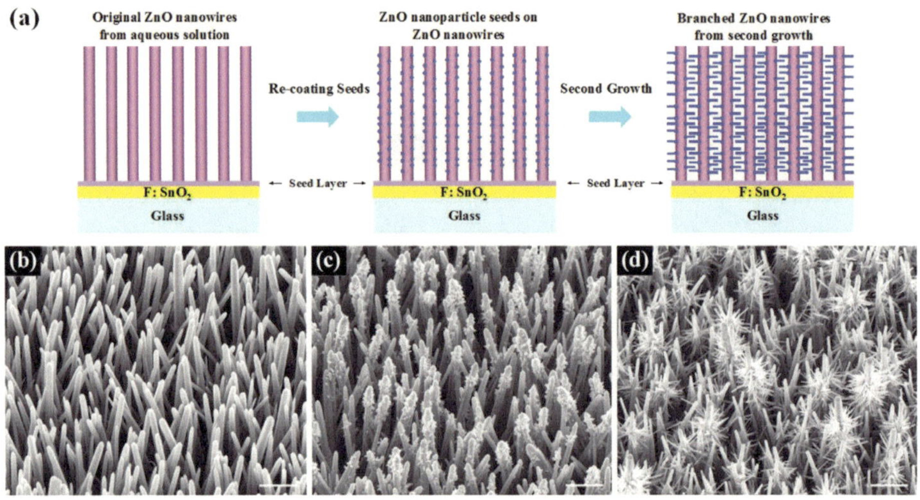

Figure 2.
(a) The schematic growth procedure from the original ZnO nanowires to the branched ZnO nanowires. (b) Before and (c) after re-coating a seed layer of the original ZnO nanowires obtained from a solvothermal method. (d) The branched ZnO nanowires after second growth. Scale bar, 1 μm [4].

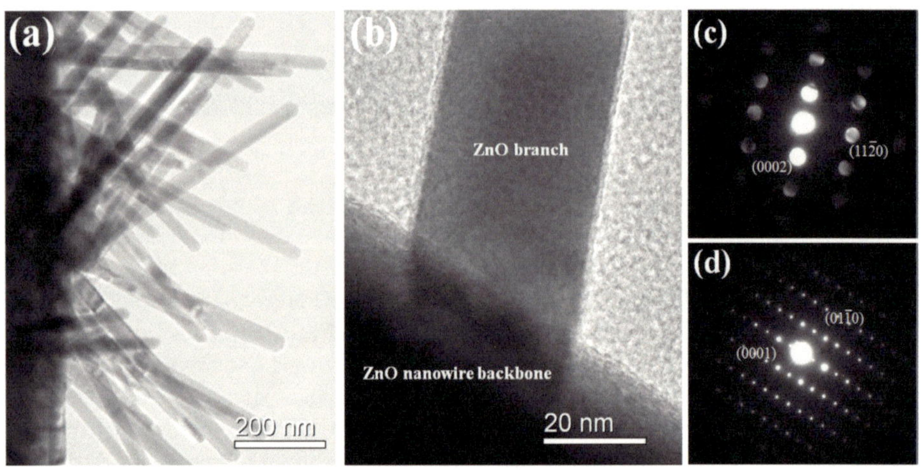

Figure 3.
(a) TEM image of a single-branched ZnO nanowire. (b) The magnified intersection area of ZnO branch and nanowire. (c and d) The corresponding nano-beam diffraction (NBD) and selected area electron diffraction (SAED) for the secondary ZnO branch and the ZnO nanowire backbone, respectively [4].

branch and nanowire was shown in **Figure 3b**. **Figure 3c**, **d** shows the corresponding nano-beam diffraction (NBD) of the secondary ZnO branch and selected area electron diffraction (SAED) of the ZnO nanowire backbone, respectively. The diffraction patterns confirmed each ZnO nanostructure was single-crystal wurtzite and preferentially oriented in the *c*-axis direction even though the two components of ZnO nanostructures were not fabricated simultaneously. The θ-2θ X-ray diffraction patterns of ZnO nanostructures, which corresponds to the hexagonal wurtzite crystallites with cell constants of a = 3.251 Å and c = 5.208 Å, were shown in **Figure 4a**. The strong {0001} diffraction family of ZnO once indicates that the nanowires are moderately oriented in the *c*-axis direction. It is presumed that all the precursors have been completely decomposed as no excess peaks can be detected. A Raman spectrum of the branched ZnO NWs, as shown in **Figure 4b**, which was taken from a 5 μm^2 spot size excited by a frequency-doubled

Yb:YAG laser (λ = 515 nm), obviously indicates the remarkable E_2 (low) and E_2 (high) modes of ZnO located at 98 and 438 cm^{-1}, respectively. The peak at 332 cm^{-1} is attributable to the second-order Raman scattering caused by the zone-boundary phonons 2-E_2(M) of ZnO. The weak and almost invisible signal near 581 cm^{-1} contributes to the superposition of A_1(LO) and E_1(LO). The substrate signal did not appear due to the penetrating limitation of the 515 nm laser. The good crystalline quality of ZnO nanostructures confirmed above ensures that the photoelectrode can provide good electronic conductivity without the defect trapping within the structures.

Secondary NPs herein were synthesized via sol-gel method [7]. The detailed synthetic process is similar to that described by Seelig et al. [8]. The structural evolution of products synthesized in various aging time using 10 ml of primary supernatant was shown in a series of SEM photographs as shown in **Figure 5**. The ZnO NPs were accumulated by white seeds from the beginning shown in **Figure 5a**.

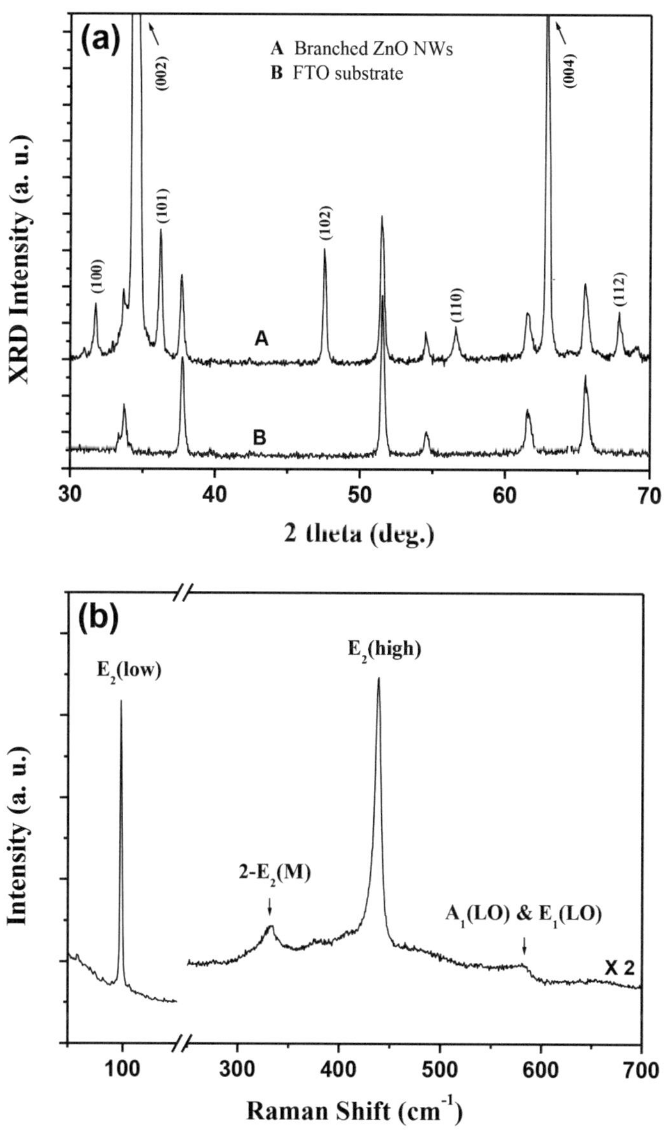

Figure 4.
(a) θ-2θ XRD profiles of A (the branched ZnO nanowires) and B (FTO substrate only). (b) Raman spectra of the branched ZnO nanowires, using a frequency-doubled Yb:YAG laser (λ= 515 nm) [4].

The zinc complexes were initially connected as a network (see **Figure 5b**) and condensed isotropic, eventually forming a hierarchical packing of colloidal particle, as shown in **Figure 5c**. The unidirectional aggregate phenomenon and formation mechanism in other metal oxide colloidal systems were presented by Serna et al. [9]. In at least 1-hour aging time, the monodispersed spherical ZnO NPs with an average particle size of ca. 185 nm could successfully be synthesized. The EDS spectra of the products with different aging time, as shown in **Figure 6**, reveal that they contain Zn, O, and C. As the aging time increases, the carbon ratio decreases, which means that the product requires complete aging time to remove the acetate ions.

Typical TEM micrographs of the ZnO NPs are shown in **Figure 7a–d**. A hierarchical packing of secondary ZnO NPs is formed in the condensation reaction of the sol-gel process, and the spherical shape of the ZnO NPs is recognized by the aggregation of many primary single crystals (also called subcrystals) ranging from 6 to 12 nm. It should be particularly noted here that when SAED is performed on several secondary ZnO NPs, the pattern exhibits a polycrystalline wurtzite structure of ZnO, as shown in the inset of **Figure 7a**. On the contrary, the pattern reveals the single-crystal-like diffraction, as shown in the inset of **Figure 7b**, while restricting the SAED area within only one ZnO NP. Obviously, the secondary ZnO NPs are polycrystals consisting of much smaller subcrystals of the same crystal orientation. More evidence can be demonstrated in the high-resolution TEM (HRTEM) image from both the center and the edge of the ZnO NPs in **Figure 7b, c**, respectively. In most cases, van der Waals interacts as a driving force for self-assembly between surface molecules of nanocrystallites, which can then assemble colloidal nanocrystals to form solids. If the size distribution of the nanocrystals is sufficiently small, an ordered array (also known as a superlattice), a quantum dot, or an artificial solid is formed by self-assembly [11, 12]. Therefore, the growth of the above secondary ZnO nanoparticles could be carried out in substantially the same way, with some discontinuities between the subunits, and each subcrystal is a subunit of the secondary ZnO NPs. Sugimoto et al. also reported similar self-assembly structures in α-Fe_2O_3 particles [13, 14]. The reason for the discontinuity of the internal structure is explained by the strong adsorption of ions used in the

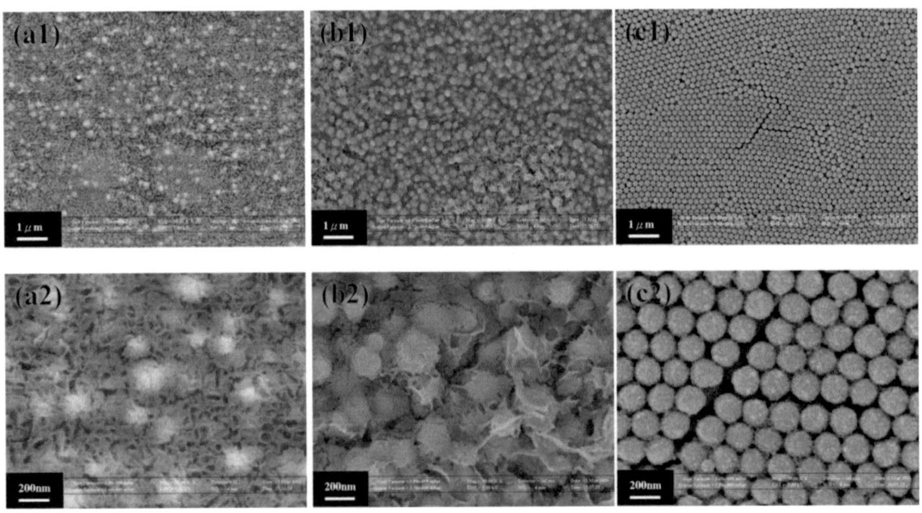

Figure 5.
Large and local scale of scanning electron micrographs of various aging time products synthesized using 10 ml of primary supernatant. The aging times are (a) 15 minutes, (b) 30 minutes, and (c) 60 minutes, respectively [7].

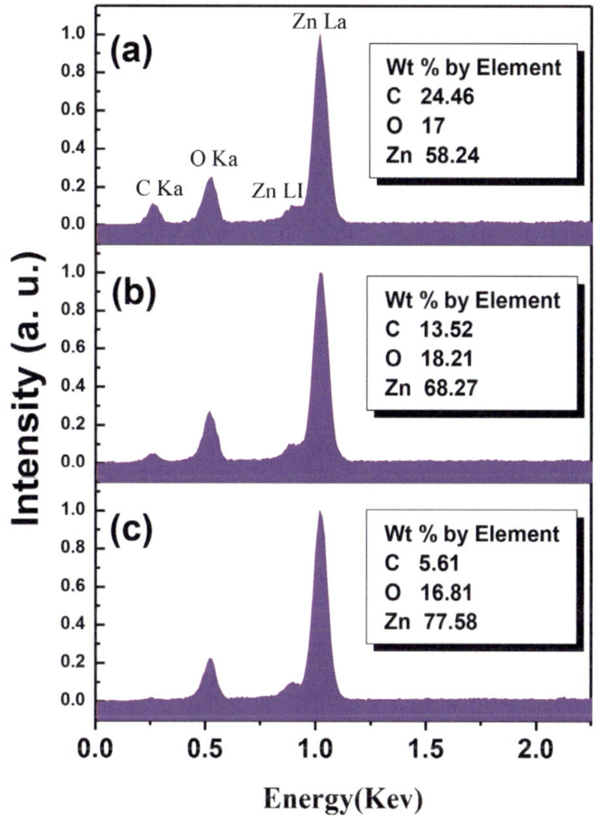

Figure 6.
Composition-variation analysis by energy dispersive x-ray spectra (EDS) of different aging time products as (a) 15 minutes, (b) 30 minutes, and (c) 60 minutes [7].

reaction to prevent the fusion between the surface grains [15–17]. Although there is no existence of ions throughout the whole process in this present work. It is suspected that due to the blockage of DEG, the solvent appears as a microemulsion system, resulting in the separate growth of the ZnO subcrystals alone and eventually assembled under the driving force of the van der Waals interaction to form secondary NPs. As shown in **Figure 7d**, the subcrystal is a perfect crystal and exhibits a facet for which the evidence is specifically described in the HRTEM image of the edge of ZnO NPs. Thus, the subcrystals slowly self-assemble by sintering and belonging to the same defined orientation as the adjacent subcrystals.

It is interesting that the evolutions of morphology of ZnO NPs show the subcrystals significantly fused with the neighbor crystals during the heating process. **Figure 8a–c** displays the SEM images of as-grown ZnO NPs and the samples after post-annealing at 350 and 500°C in air ambient for 1 hour, respectively. The grain growth was also investigated from the XRD profiles (not shown here) and Scherrer formula, as the crystalline sizes were estimated to be approximately 9, 14, and 20 nm for as-grown, 350°C-annealed, and 500°C-annealed samples, respectively. Raman spectroscopy was performed to investigate the vibrational properties of the secondary ZnO NPs before and after being heat treated. For the wurtzite structure of ZnO, which belongs to the space group $C_{6\upsilon}^{4}$ ($P6_3mc$), one primitive cell includes two formula units, with all of the atoms occupying $2b$ sites of symmetry $C_{3\nu}$. **Figure 9** shows a normal Raman spectra by a frequency-doubled Yb:YAG laser (λ = 515 nm). The remarkable feature at 520 cm^{-1} is due to the TO phonon mode from the Si substrate, while the peak at 437 cm^{-1} corresponds to E_2(high) of

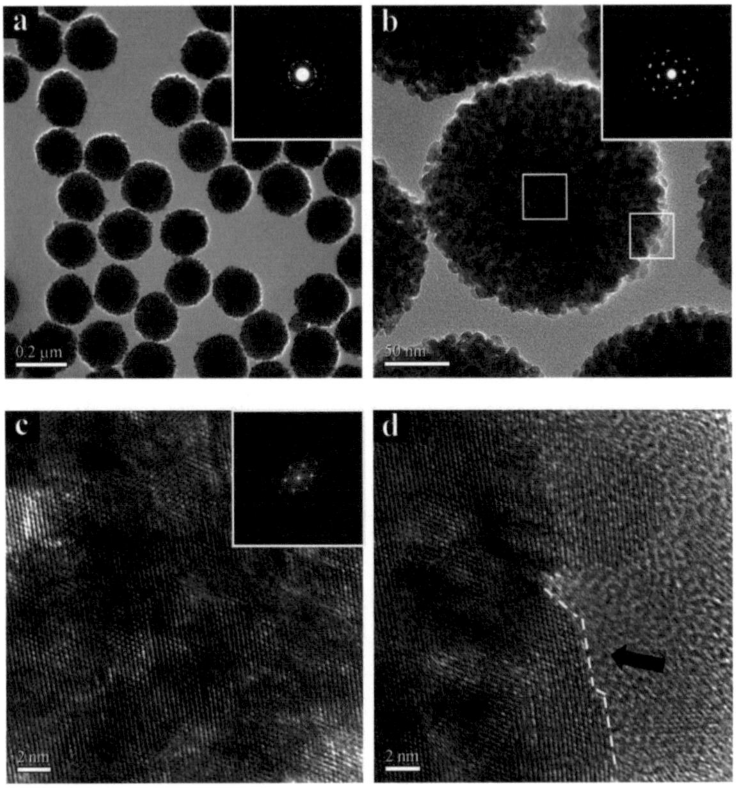

Figure 7.
TEM images of secondary ZnO NPs recognized of crystalline subcrystals. (a) A typical low-magnification TEM image and SAED pattern of several uniform ZnO NPs. (b) High-magnification TEM image of one individual ZnO NP and its corresponding single crystal-like SAED spots. (c and d) High-resolution TEM images of central area and boundary part of one individual ZnO NP, respectively. Inset of (c) corresponding fast Fourier transform image [10].

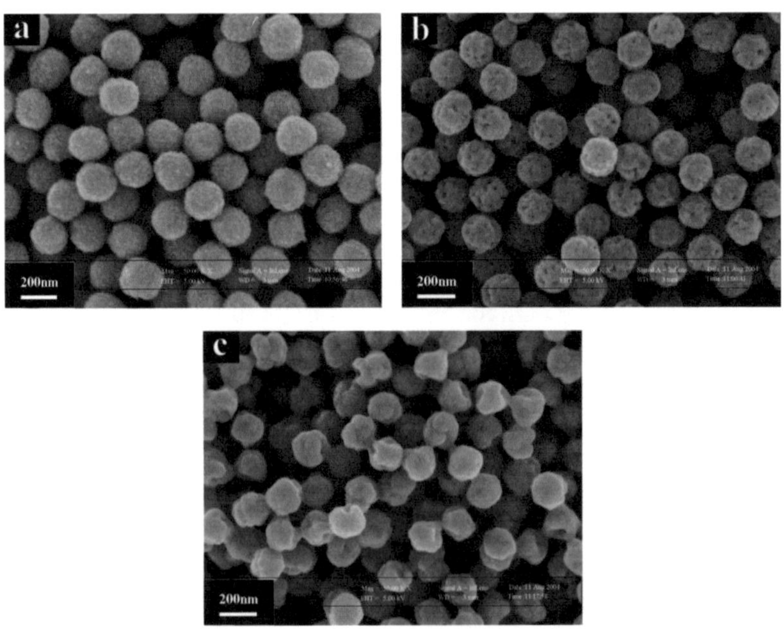

Figure 8.
SEM micrographs of (a) as-grown, (b) 350°C annealing for 1 hour, and (c) 500°C annealing for 1 hour secondary ZnO NPs, respectively [10].

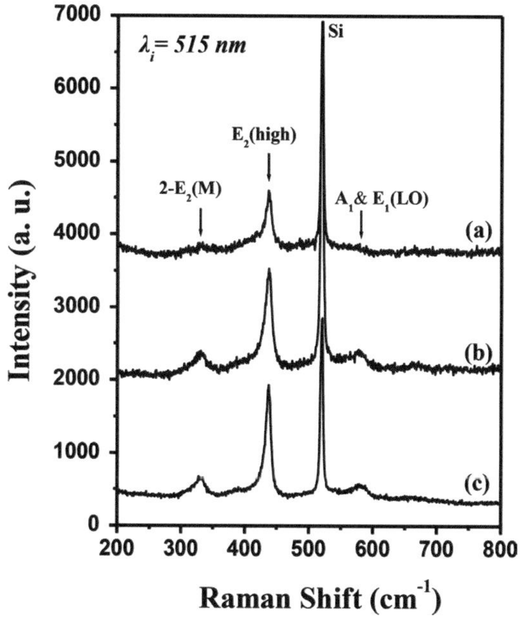

Figure 9.
Normal Raman spectra of (a) as-grown, (b) 350°C annealing for 1 hour, and (c) 500°C annealing for 1 hour secondary ZnO NPs, using a frequency-doubled Yb:YAG laser (=515 nm) [10].

ZnO. The peak at 331 cm^{-1} can be assigned to the second-order Raman scattering arising from zone-boundary phonons 2-E_2 (M) of ZnO. No significant change of Raman spectra and intense E_2 (high) peak for the 350°C-annealed and 500°C-annealed samples means good crystallinity. The full width at half-maximum (FWHM) of Raman E_2(high) peak decreases (from 14 to 11 cm^{-1}) as the crystal size increases as anneal temperature from 350°C to 500°C, which is consistent with the XRD results. Another imperceptible broadened peak around 580 cm^{-1} is contributed to the superposition of A_1(LO) and E_1 (LO). The lattice behavior of A_1 (LO) and E_1 (LO) modes is associated with the existence of some nonstoichiometric defects while heat treatment, such as oxygen vacancy, interstitial zinc, or their complexes [18–20] those are produced due to the unfavorable process environment.

In principle the electron-phonon interaction could be investigated by utilizing resonant Raman scattering (RRS) experiments. The excitation photon energy resonates with the transition energy above the electron bands of wurtzite ZnO, so that a He-Cd laser (= 325 nm) was used as the excitation source for RRS. The A_1 (LO) and E_1 (LO) modes would dominate as their polar symmetric and exhibit different frequencies from the TO modes as well. As shown in **Figure 10**, intense multiphonon scatterings of the secondary ZnO NPs before and after heat treatment were observed, where the major peaks were observed as a result from the polar symmetry modes A_1 (LO) and E_1(LO) and their overtones. For the superposition of LO phonon mode, the Zn atoms and the O atoms have the same vibration direction as the adjacent lattice, respectively [21], while the weak peak, which is contributed to the E_2(high) mode around 437 cm^{-1}, is almost imperceptible. Multiphonon scattering processes also have been previously reported for single-crystalline bulk ZnO [22], ZnO films [23], ZnO-opal structures [24], and ZnO NWs [25, 26] but rarely mentioned for ZnO NPs.

It is worth noting that the intensity of the first-order Raman mode and its overtone are enhanced in the grown ZnO NP compared to the annealed sample. The reason can be explained by the total Raman cross section for an n-phonon process written as [27, 28]

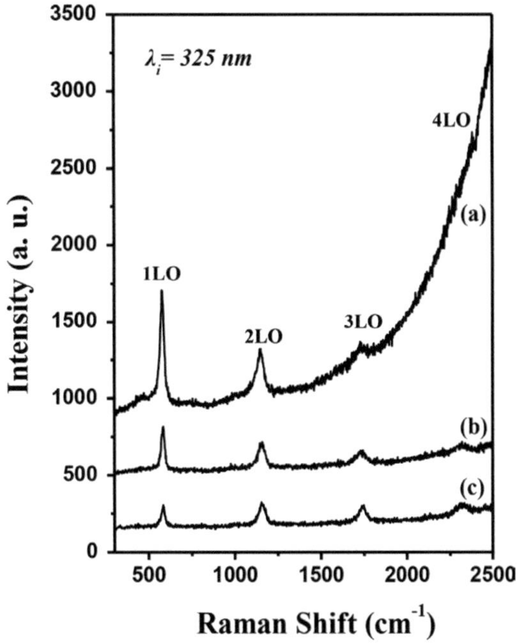

Figure 10.
Resonant Raman scatterings (RRS) of (a) as-grown, (b) 350°C annealing for 1 hour, and (c) 500°C annealing for 1 hour secondary ZnO NPs, using a He-Cd laser (= 325 nm) [10].

$$\sigma_n = \int \sigma_n^R(\omega) f(R) dR, \qquad (1)$$

$$\sigma_n^R(\omega) = \mu^4 \left| \sum_{m=0}^{\infty} \frac{\langle n|m\rangle\langle m|0\rangle}{E_0 + n\hbar\omega_{LO} - \hbar\omega + i\hbar\Gamma} \right|^2 \times \exp\left(-\frac{i\hbar\omega_{LO}}{k_B T}\right), \qquad (2)$$

where μ is the electronic dipole transition moment; E_0 is the size-dependent energy of the electronic transition; $\hbar\omega$ and $\hbar\omega_{LO}$ are the energies of the excitation photon and the LO phonon, respectively; m denotes the intermediate vibrational level in the excited state; Γ is the homogeneous linewidth; k_B is Boltzmann's constant; T is the temperature; and the bracket indicates the overlap integral between the ground and excited state wave functions. Consequently, the RRS intensity can be enhanced as the denominator in Raman scattering cross section tending to zero, while the electronic state in the material is close to the incident or scattered photons. Similar results have been previously reported for CdS and ZnO, using various laser wavelengths [29]. Due to the quantum confinement effect of the subcrystal size relative to the exciton radius, the bandgap of the as-grown ZnO NPs would tend to approach the excitation laser energy. Evidence of quantum confinement can be found by the intensive tail of the blue-shifted photoluminescence (PL) signal of the as-grown ZnO NPs, or it can be found in the RRS spectrum rather than in the normal RS spectrum due to their red-shifted, broadening and asymmetry.

Due to the infinite correlation length, the phonon eigenstate in an ideal crystal is a plane wave; therefore, the $K = 0$ momentum selection rule of the first-order Raman spectrum can be satisfied. When the crystalline is reduced to nanometer scale, the momentum selection rule will be relaxed. This allows the phonon with wave vector $|k| = |k'| \pm 2\pi/L$ to participate in the first-order Raman scattering, where k' is the wave vector of the incident light and L is the size of the crystal.

The phonon scattering is not limited to the center of the Brillouin region. In order to observe the displacement, broadening, and asymmetry of the first-order optical phonon, the phonon dispersion near the center of the region must also be considered. Alim et al. [30, 31] have shown that the large red-shift in the resonant Raman spectrum from 20 nm ZnO NPs is most likely due to local heating by UV laser excitation. In this study, since the as-grown secondary ZnO NPs contain more air gaps than the annealed NPs, an unfavorable heat dissipation may be the other possibility of causing higher temperatures and greater phonon red-shift. For the clarity, detailed numerical analysis of the ZnO NPs of this experiment is clearly listed in **Table 1**. It was found that the ratio between the second-order and first-order Raman scattering cross sections increased from 0.38 to 2.05, while the ZnO crystallite size increased from nanoparticle to bulk. In the Franck-Condon approximation, the coupling strength of the [32, 33] exciton transition to the LO phonon can be expressed by the Huang-Rhys parameter S. The cross section of the RRS depends on the particle size, temperature, and excitation wavelength. Scamarico et al. [34] proposed that due to the strong energy dependence of the Raman scattering cross section, it is necessary to maintain resonance conditions in order to make meaningful comparisons with spectra of nanocrystals of different sizes to maintain different electronic transitions. It is important to purposefully use the same experimental conditions, such as laser power, wavelength, spot size, etc. for each sample. In this study, I emphasize that the tendency here is the increasing electron-phonon interaction with increasing nanocrystal size. It is generally accepted that the electron-phonon coupling is determined by the deformation potential and the Fröhlich potential. TO Raman scattering cross section is mainly determined by the deformation potential that involves the short-range interaction between the lattice displacement and the electrons [35, 36]. On the other hand, the LO Raman scattering cross section includes contributions not only the Fröhlich potential that involves the long-range interaction generated by the macroscopic electric field associated with the LO phonons but also the deformation potential. The intensity of TO phonons in ZnO NPs was found to be almost insensitive, while the intensity of LO phonons was greatly enhanced under resonance conditions. This study shows that electron-LO-phonon coupling is related to the Fröhlich interaction as the size of the nanocrystals decreases. Although the complex origin is not clear, the results of this study are very similar to those of other low-dimensional ZnO nanostructures, such as ZnO-based quantum wells [37] and ZnO NWs [38].

	As-grown[a]	**350°C**[a]	**500°C**[a]	**Bulk**[b]
Grain size (nm)	9	14	20	>1000
1LO (cm^{-1}) (FWHM)	578 (27.2)	582 (24.2)	584 (23.9)	585 (N/A)
2LO (cm^{-1}) (FWHM)	1149 (54.7)	1154 (49.3)	1158 (47.3)	1165 (N/A)
I_{2LO}/I_{1LO}	0.38	0.59	1.07	2.05

[a]*This content*
[b]*From Ref. [22] (used the same He-Cd laser, λ= 325 nm, as the excitation source for RRS)*

Table 1.
Wave number, broadening, and the ratio of n-LO phonons found in RRS spectra. The assignments of bulk ZnO are also listed as a Ref. [10].

3. ZnO nanostructures for dye-sensitized solar cells

DSCs were prepared by immersing the area-defined ZnO NW specimens into a solution of 0.5 mM cis-bis(isothiocyanato)bis(2,2′-bipyridyl-4,4′-dicarboxylato)-ruthenium(II)bis-tetrabutylammonium (N719, Solaronix) in acetonitrile/*tert*-butanol (1:1) for 20 minutes. After rinsing with acetonitrile and natural drying, the sensitized electrodes were sandwiched together with thermally platinized FTO counter electrodes separated by 25-μm-thick hot-melt spacers (Surlyn, Dupont). The electrolyte solution (0.1 M LiI, 0.5 M 1,2-dimethyl-3-propylimidazolium iodide, 0.03 M I_2, and 0.5 M *tert*-butylpyridine in acetonitrile) was filled into the internal space.

The comparison of photocurrent-voltage (J-V) characteristics for solar cells which was constructed using the bare ZnO NWs and the branched ZnO NWs with AM 1.5 illumination at 100 mW/cm^2 from a xenon lamp was shown in **Figure 11**. The overall light conversion efficiency and short-circuit current density (J_{sc}) of the branched ZnO nanowire DSCs were 1.51% and 4.27 mA/cm^2, respectively, which are almost twice higher than that of the bare ZnO NWs. Increased photon absorption is associated with an increase in internal surface area, resulting in enough dye loading as a major factor in the increase in short-circuit current density. Although the density of ZnO structures exhibits the insignificancy compared with some previous studies, the shortage can be complemented via the extra branches. The values of fill factor (*FF*) for ZnO DSCs are generally low (~0.5) which is attributed to recombination between photoexcited carriers and triiodide ions in the photoanodes and electrolyte, respectively. No obvious difference of the shunt resistance $R_{sh} = (dV/dI)_{V=0}$ from the J-V curves under illumination revealed almost the same interfacial recombination; however, the series resistance $R_s = (dV/dI)_{I=0}$ for branched ZnO nanowire DSCs (25.64 Ωcm^2) was significantly lower than the bare ZnO nanowire ones (46.13 Ωcm^2).

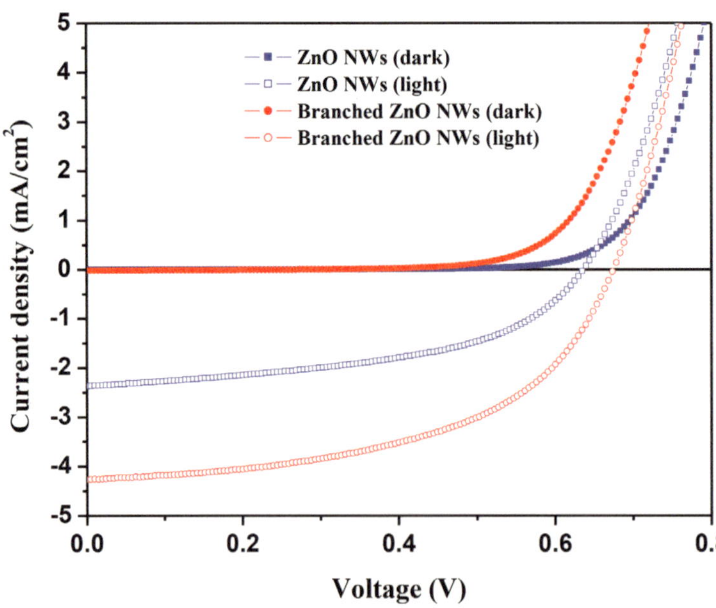

Figure 11.
Current density against voltage (J-V) characteristics of the bare ZnO nanowires and the branched ZnO nanowire DSCs [4].

Some interior parameters of DSCs can be analyzed by the impedance data of the Nyquist plots. The modified equivalent circuit of nanowire DSCs was suggested by Wu et al. [39, 40]. **Figure 12** shows the impedance data for bare and branched ZnO nanowire DSCs performed by applying a 10 mV ac signal over the frequency range of 10^{-2}–10^{5} Hz under illumination at the applied bias of V_{oc}. r_w and r_k are the transport resistance of the electrons in the ZnO electrode and the charge-transfer resistance of the charge recombination between electrons in the ZnO electrode and I_3^- in the electrolyte, respectively. The thickness L_F of all anodes are about 8 μm; $C_\mu = (c_\mu L_F)$ is the chemical capacitance of the ZnO electrode; Rs is a summary resistance for the transport resistance of FTO and other resistances out of the cell; Z_N is the impedance of diffusion of I_3^- in the electrolyte; C_{Pt} and R_{Pt} are the interfacial capacitance and the charge-transfer resistance at the counter electrode (platinized FTO glass)/electrolyte interface, respectively; C_{FTO} and R_{FTO} are the interfacial

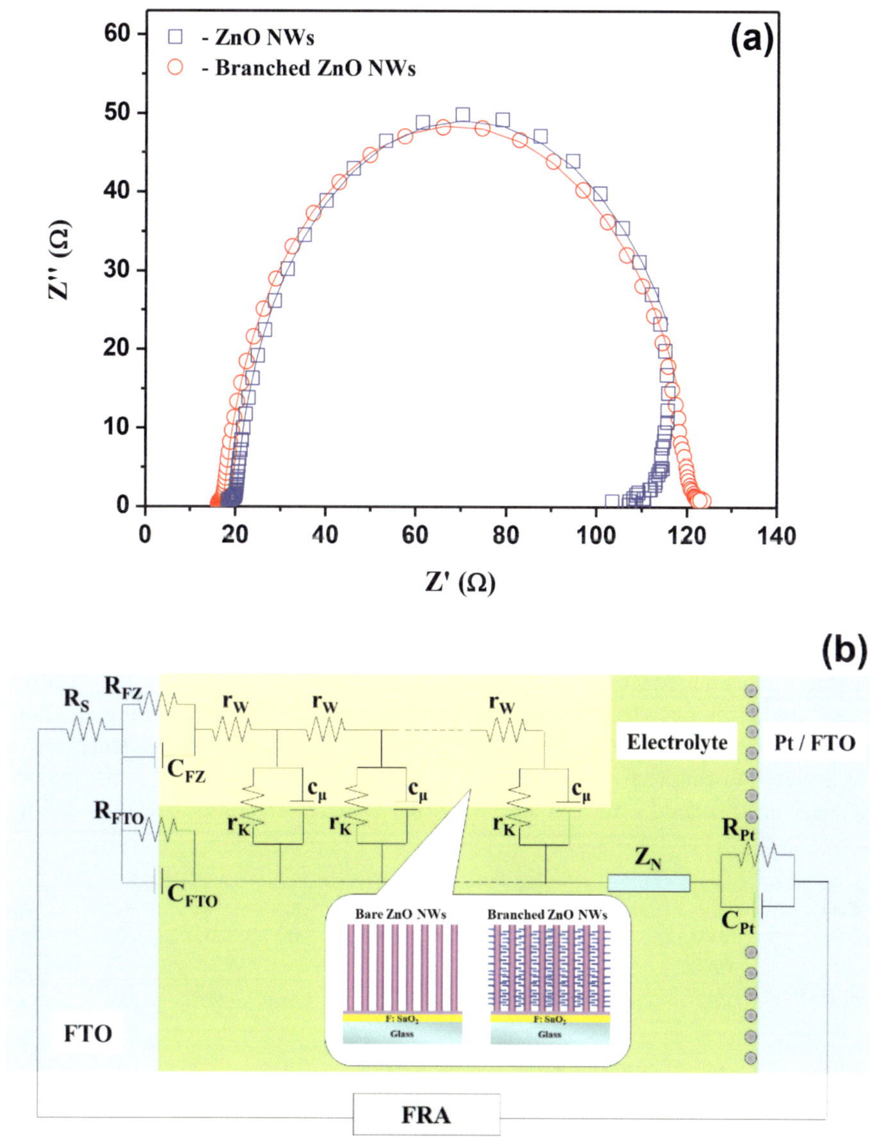

Figure 12.
Nyquist plots of the bare ZnO nanowires and the branched ZnO nanowire DSCs. The solid lines are the fitting results based on the equivalent circuit model (modified from ref. [41] as shown in the inset) [4].

capacitance and the charge-transfer resistance at the exposed FTO/electrolyte interface, respectively; C_{FZ} and R_{FZ} are the capacitance and resistance at the FTO/ZnO contact, respectively.

The specific equivalent circuit might be more complicated while operating the DSCs. Mora-Seró et al. have reported the conductivity modulation of the electrolyte-induced negative capacitance which comes from by injected electrons from the photoelectrode while providing high forward bias with low frequency [42]. In order to avoid the unnecessary interference from the inductor, the low-frequency part of impedance data was currently ignored. The fitted results of the first-order reaction rate constant for the loss of electrons(k_{eff}), the electron lifetime ($\tau = 1/k_{eff}$), the electron transport resistance ($R_w = r_w L_F$), and the charge-transfer resistance related to recombination of an electron at the ZnO/electrolyte interface ($R_k = r_k/L_F$) were listed in **Table 2**. R_k and R_w are quite similar for both DSCs which meant the same crystallinity and interfacial recombination for either bare ZnO NWs or branched ZnO nanostructures. k_{eff} in the branched ZnO nanowire DSCs was smaller than the bare nanowire ones to cause the smaller effective diffusion length [43] ($D_{eff} = (R_k/R_w)L_F^2 k_{eff}$) in branched ZnO nanowire DSCs. But the electron lifetime ($\tau_{eff} = 1/k_{eff}$) was prolonged by the additional transport distance of branched ZnO nanowire DSCs.

The current density of a DSC is determined by the amount of photogenerated carriers, the electron injection efficiency from the dye molecules to the semiconductor, and the recombination rate between the injected electrons and the oxidative dye or redox species in the electrolyte. The initial amount of photogenerated carriers may have a significant effect on the light-harvesting ability of photoanodes of different structures. **Figure 13** displays the comparison of incident monochromatic photon to current conversion efficiency (IPCE). The peaks at approximately 400 nm were due to direct light harvesting by ZnO semiconductor. The photogenerated electrons diffused through ZnO and the holes in the valence band were replenished directly by charge transfer from the I_3^-/I^-electrolyte [44]. The maximum peak at approximately 525 nm is contributed by the dye absorption, corresponding to the visible $t_2 \rightarrow \pi^*$ metal-to-ligand charge transfer (MLCT). The IPCE obtained for the branched ZnO nanowire DSCs was almost 1.5 times that of the bare ones. This improvement is primarily due to sufficient dye loading of the branched ZnO NWs, which increases the internal surface area within the photoelectrode. From the dye loading measurement, as shown in **Figure 14**, the concentration of dye in the branched ZnO nanowire electrode was found to be 2.9×10^{-9} mol/cm^{-2} as measured from dye-desorption experiments, which is almost 40% higher than the obtained value of 2.1×10^{-9} mol/cm^{-2} for the bare ZnO nanowire electrode.

It is worth noting that even though the current density and energy conversion efficiency of branched ZnO DSCs is twice that of bare ZnO ones, the dye loading

ZnO DSCs	J_{sc} (mA/cm^2)	V_{oc} (V)	*FF*	η (%)	k_{eff} (s^{-1})	τ_{eff} (s)	R_k (Ω)	R_w (Ω)	D_{eff} (cm^2/s)
Bare nanowires	2.37	0.636	0.498	0.75	38.31	0.026	92.12	3.63	6.23×10^{-4}
Branched nanowires	4.27	0.675	0.522	1.51	26.31	0.038	86.85	3.36	4.35×10^{-4}

Table 2.
Performances and electron transport properties of the bare ZnO nanowire and the branched ZnO nanowire DSCs determined by photocurrent density-voltage (J-V) characteristics and electrochemical impedance spectroscopy (EIS) analysis [4].

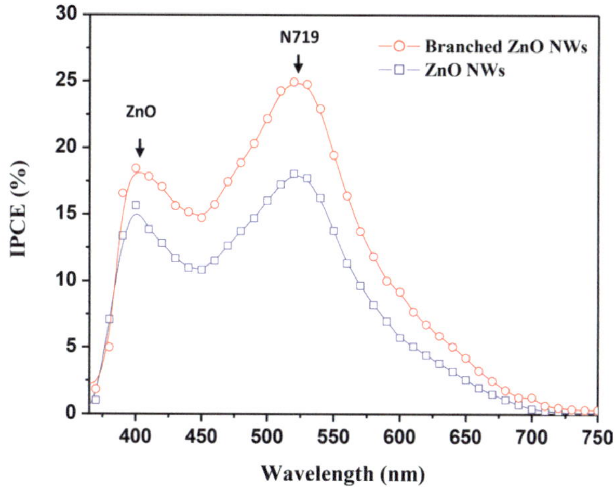

Figure 13.
The incident monochromatic photon to current conversion efficiency (IPCE) of the bare ZnO nanowire and the branched ZnO nanowire DSCs [4].

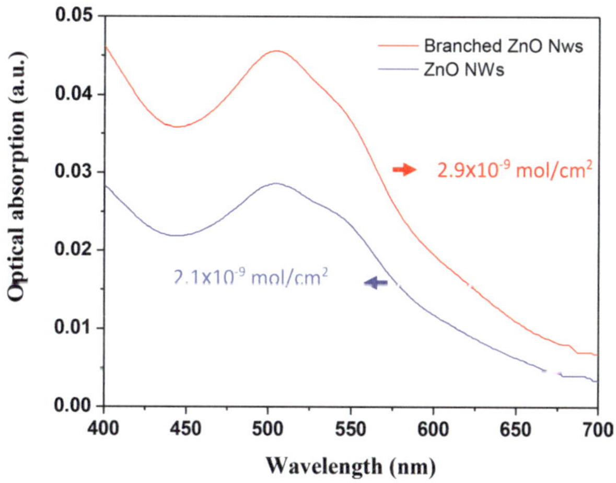

Figure 14.
Optical absorption of dye detached from the bare ZnO nanowire and the branched ZnO nanowire substrates and dissolved in 0.1 M NaOH solution [4].

of branched ZnO DSCs is not twice that of bare ZnO ones. The differences may be that the internal surface of the bare ZnO NWs is insufficient and the excess dye results in the formation of Zn^{2+}/dye complexes rather than an effective chemical bond between the ZnO and dye molecules. To avoid the excessive reaction, Chou et al. [45] reported the shorter immersion time for deficient chemical stability ZnO electrode as compared with TiO_2 electrode. Therefore, the excess immersion time of the insufficient internal surface of bare ZnO NWs cannot achieve more dye loading but seriously deteriorates the performance of the DSCs. Although the secondary branches have a nonuniform distribution due to the simple dip-coating process, the branches emitted from a portion of the ZnO NWs could still provide a greater effective surface area for dye adsorption than the bare NWs. The dc or radio frequency (RF) magnetron sputtering or atomic layer deposition (ALD) is suggested to apply great benefit to the pre-coating processes for the optimization of the filling factor of the ZnO NWs. Further improvement of light harvesting, current density, and overwhelm energy conversion efficiency could be implemented through adjusting denser and longer branches to fill the interstitial voids between backbone NWs.

Hierarchically packed ZnO NPs were formed in the condensation reactions of the sol-gel process that was mentioned previously. The diameter of the ZnO NPs in the range of 160–680 nm was used as shown in **Figure 15a**. **Figure 15b** shows the spherical-shaped secondary ZnO NPs with the diameter of 680 nm. The similar ZnO architectures have been elucidated as the random lasing applications in which the cavities were formed by multiple scattering (UV range) between ZnO primary particles [47]. The laser action comes from an efficient amplification along the closed-loop light-scattering path within a secondary ZnO nanoparticle. Recently, Cao et al. have demonstrated that the aggregation of ZnO nanocrystallites performs an effective scheme to generate multiple light scattering (sunlight range) within the photoelectrode film of DSCs without using any other scattering layers [48, 49], as shown in **Figure 15c**. With utilization of ruthenium complex *cis*-[$RuL_2(NCS)_2$] (L = 4,4′-dicarboxy-2,2′-bipyridine), N3 dye, Cao achieved the maximum energy conversion efficiency of 5.4%.

In this present research, a broad size distribution of secondary ZnO NPs with mean radius of 360 nm, as shown in **Figure 16a**, is controlled to provide the wide-range absorption of visible sunlight within the preferable packing of the ZnO photoelectrode. The hierarchical ZnO photoelectrode provides the multiple scattering of light, and therefore the light-traveling distance can be significantly prolonged. Furthermore, the primary ZnO nanocrystallites could supply internal surface area to allow enough adsorption of dye molecules. From the optical absorption spectra, the intrinsic exciton absorption (direct transition of energy bandgap of ZnO) could

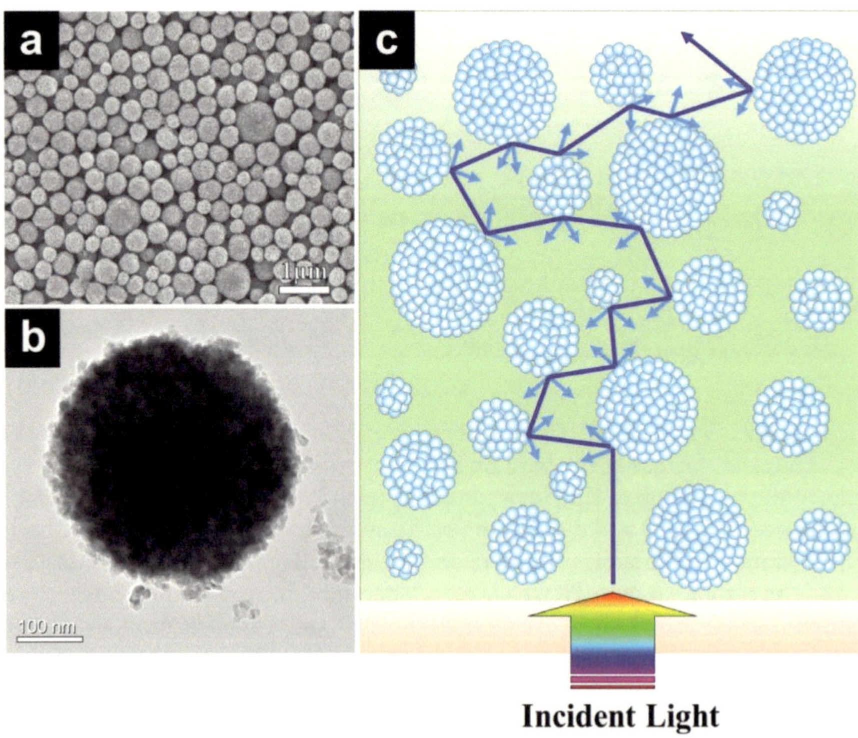

Figure 15.
(a, b) The FESEM and TEM images for the self-assembled ZnO secondary nanoparticles, respectively. (c) The schematic multiple scattering of light within the hierarchical ZnO photoelectrode composed by self-assembled ZnO secondary nanoparticles [46].

be particularly identified for 2 μm film. However, the absorption at wavelengths around 400–650 nm caused by the light scattering is enhanced dramatically with increasing the thickness of the ZnO photoelectrodes from 2 to 12μmm as shown in **Figure 16b**. Through a significant light scattering from the hierarchical structure, the thicker ZnO films provide the optical gallery that could provide more photon absorption in the visible region by the dye molecules.

The fabrication procedure of DSCs for ZnO NPs were similar to the ZnO NWs, but the photoelectrodes were prepared by screen-printing method. The molecular structures of the indoline-based organic dyes employed in this hierarchically packed ZnO photoelectrode are depicted in **Figure 17**. Both D149 and D205 sensitizers have double rhodanic acid as an anchor moiety. However, D205 is designed by introducing an octyl substitute into the terminal rhodanine ring to replace the ethyl group of D149 [51, 52]. In order to improve the DSCs performance, optimization of the thickness of the ZnO photoelectrode is necessary, because the photovoltaic characteristics exhibit significant variation depending on the thickness. **Figure 18a, b**

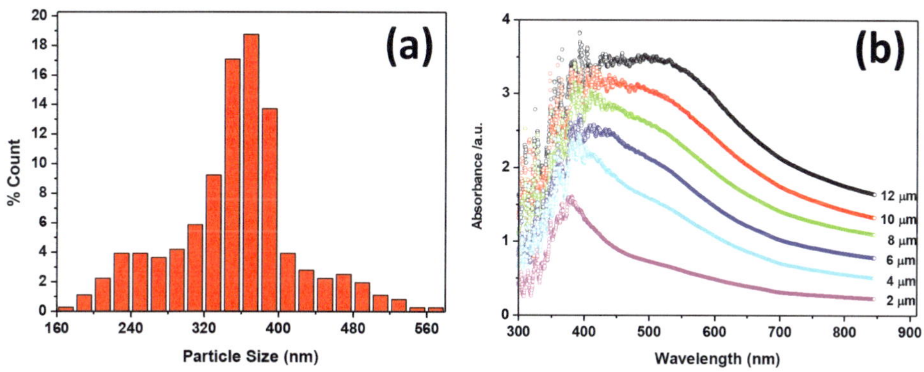

Figure 16.
(a) Diameter distribution for the ZnO secondary nanoparticles. (b) The corresponding optical absorption spectra of ZnO photoelectrodes with various film thicknesses, from 2 to 12 μm [50].

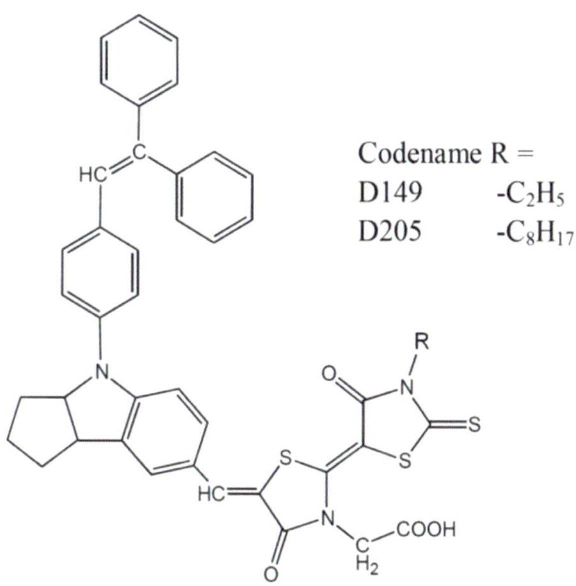

Figure 17.
Molecular structures of indoline D149 and D205 dyes.

compares the IPCE spectra of ZnO DSCs constructed using two indoline dyes with different film thicknesses. For both two indoline DSCs, IPCE increases significantly with thicker photoelectrodes but saturated at the thickness above 30 μm due to the limitation of electron diffusion length. The spectra at wavelengths shorter than 400 nm are deteriorated due to the UV cutoff effect caused by the thick glass substrate. The photocurrent peak at approximately 367 nm belongs to direct light harvesting of ZnO semiconductor, which remains almost unchanged due to the short penetration depth of UV light. **Figure 18a**, **b** also shows the maximal IPCE value increase gradually with the thickness of ZnO photoelectrode from 71 to 74% and 77 to 79% at wavelength of 550 nm for D149- and D205-sensitized DSCs, respectively. The optimal IPCE obtained for the D205-sensitized ZnO DSCs are

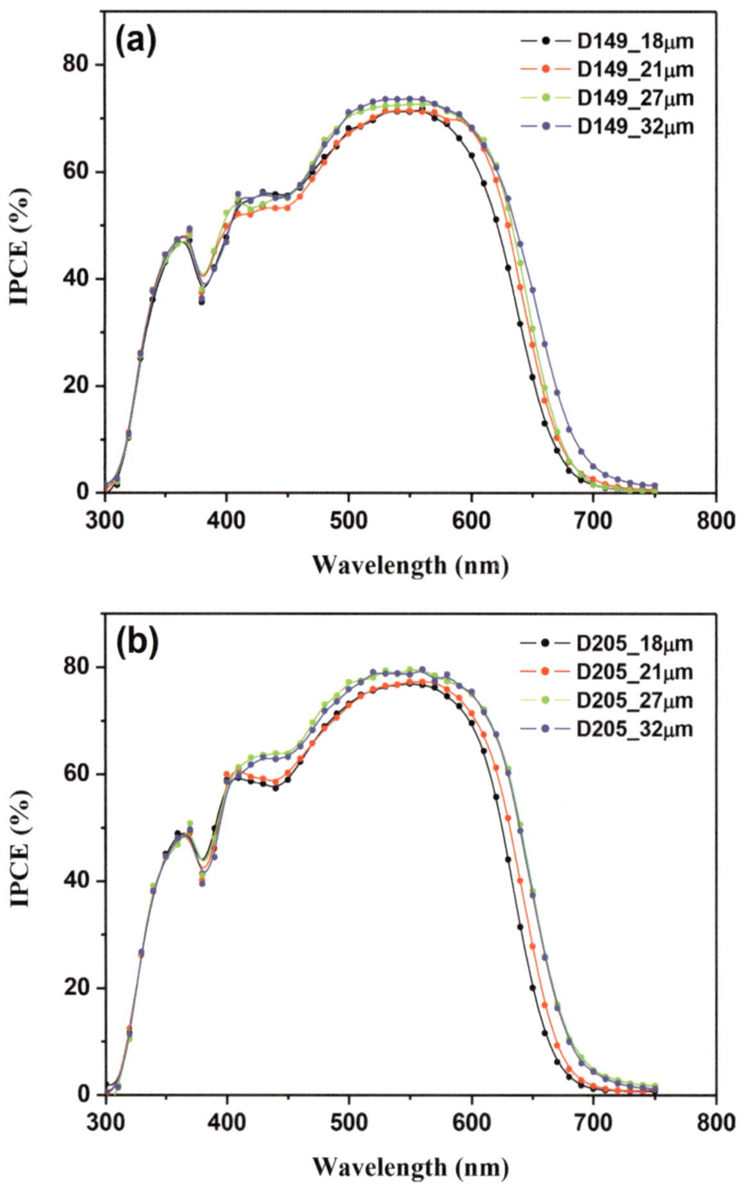

Figure 18.
Photocurrent action spectra of ZnO DSCs constructed using (a) D149 and (b) D205, with different photoelectrode thicknesses [50].

higher than the D149-sensitized DSCs in the visible-wavelength (400–700 nm) region. The numerous cracks in thick photoelectrode films (>32 μm) were also observed due to unpracticed-printing technique.

The comparison of photocurrent-voltage (J-V) of DSCs using 27-μm-thick ZnO photoelectrodes and two indoline dyes under AM 1.5 full sunlight illumination (100 mW cm^{-2}) and in the dark was shown in **Figure 19a**. For D205 uptake, the J-V plot reveals J_{sc} = 12.17 mA cm^{-2}, V_{oc} = 0.65 V, FF = 0.67, and η = 5.34%. The J-V plot of D149 uptake reveals J_{sc} = 10.94 mA cm^{-2}, V_{oc} = 0.64 V, FF = 0.71, and η = 4.95%. Both V_{oc} and J_{sc} for D205-sensitized ZnO DSCs are higher than D149-sensitized ones. As a result of the higher IPCE, the J_{sc} for the D205-sensitized ZnO DSCs is higher than the D149-sensitized ZnO ones. The dark current indicates that D205-sensitized ZnO DSCs have a slightly more negative-onset potential for the reduction of I_3^- than

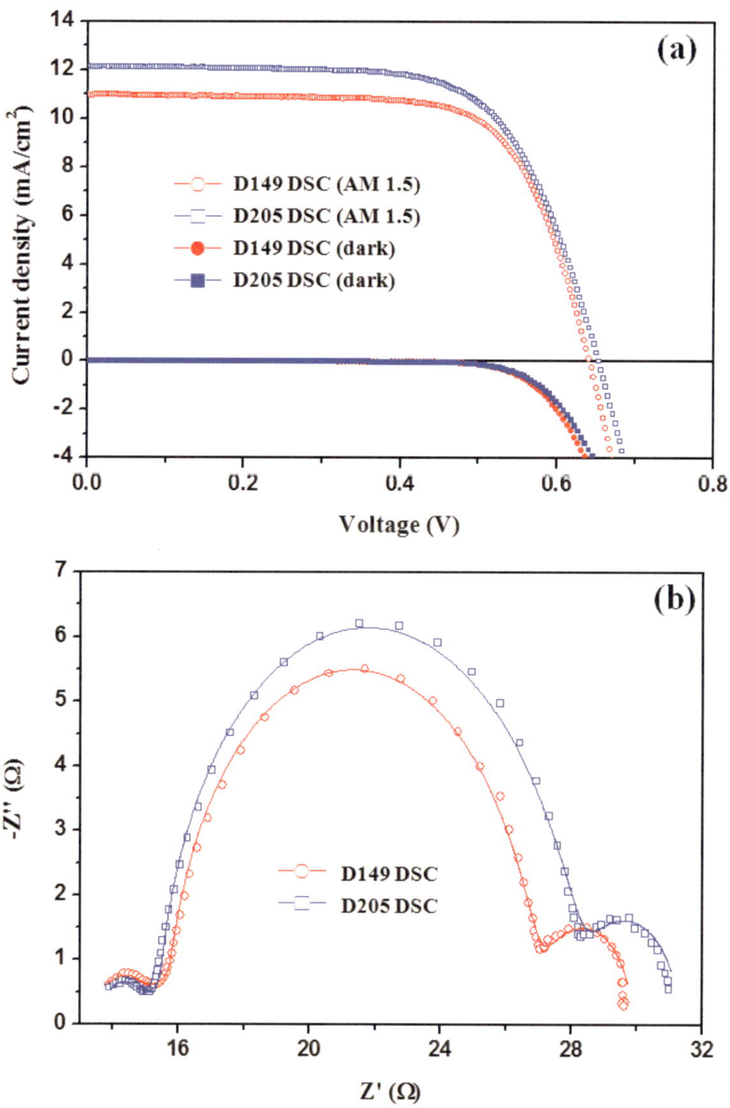

Figure 19.
Photovoltaic characteristics of DSCs with 27-μm-thick ZnO photoelectrodes and two different indoline dyes. (a) J-V curves for D149- and D205-sensitized DSCs with AM 1.5 illumination and in the dark, respectively. (b) Nyquist plots of D149- and D205-sensitized DSCs performed under illumination at the applied bias of V_{oc}. The solid lines are the fitting results [50].

D149-sensitized ZnO DSCs and could also be rationalized as a negative shift in edge of ZnO conduction band caused by D205 dye adsorption. The alkyl chain of the terminal rhodanine moiety of D205 has been extended from ethyl to octyl, which cause the effective suppression of electron recombination between I_3^- and efficient electrons injection into the photoelectrodes [53]. The dyes with the hydrophobic alkyl chains principally not only could form a barrier layer on the sensitizer dye to protect the dye layer against water intrusion from the electrolyte but also rearrange dyes that are more perpendicular to the ZnO surface.

As shown in **Figure 19b**, the obvious middle semicircle of the Nyquist plots belongs to the electron recombination resistance which means a superior circuit in D205-sensitized ZnO DSCs than that of D149-sensitized ZnO DSC. Like **Figure 12**, some interior parameters of the devices can be further derived by well-fitting the impedance data based on the modified equivalent circuit of DSCs as shown in **Figure 20**. The detail parameters are listed in **Table 3**. The electron loss rate k_{eff} in the D205-sensitized ZnO DSCs is smaller than the D149-sensitized ones, which causes the prolonged electron lifetime τ_{eff} in the D205-sensitized ZnO DSCs. The larger charge-transfer resistance R_k value for D205-sensitized ZnO DSCs indicates the less interfacial recombination occurring between the injected electrons of ZnO and the I_3^- of electrolyte. Moreover, the effective electron diffusion coefficient D_{eff} is also enhanced with utilization of D205 sensitizer. It is reasonable that the photocurrent density may be directly affected by changes in the electron recombination rate. The amphiphilic D205 may help the formation of a self-assembled dye monolayer that prevents photoelectrons from being resorbed by triiodide ions in the electrolyte, resulting in a higher V_{oc} and J_{sc} [54].

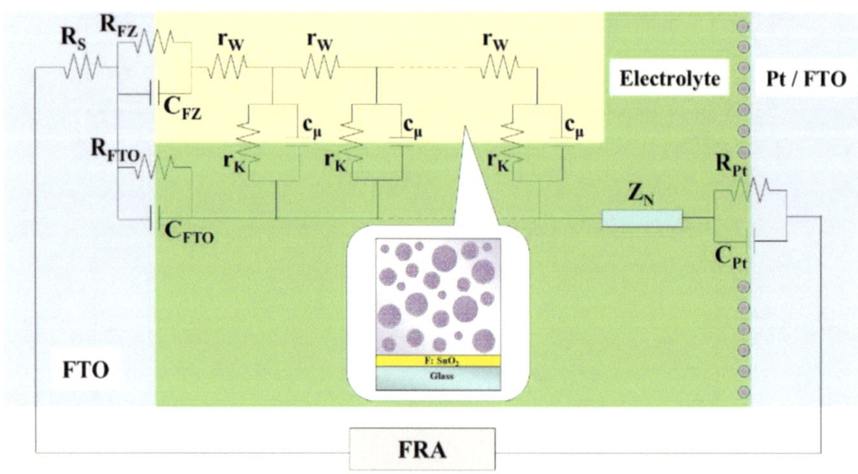

Figure 20.
The equivalent circuit model of ZnO DSCs composed with hierarchical nanoparticles [50].

ZnO DSCs	J_{sc} (mA/cm^2)	V_{oc} (V)	FF	η (%)	k_{eff} (s^{-1})	τ_{eff} (ms)	R_k (Ω)	R_w (Ω)	D_{eff} (cm^2/s)
D149	10.94	0.641	0.71	4.95	57.85	17.29	12.55	2.47	2.14×10^{-3}
D205	12.17	0.653	0.67	5.34	47.12	21.22	14.43	1.98	2.51×10^{-3}

Table 3.
Performances and electron transport properties of the D149- and D205-sensitized DSCs (27-μm-thick ZnO photoelectrode) determined by J-V characteristics and EIS analysis [50].

DOI: http://dx.doi.org/10.5772/intechopen.85699

4. Conclusion

Low-dimensional NPs, QDs, and NWs have attracted considerable attention owing to their interesting physical and chemical properties. ZnO NWs can shed the light to conduct electronic, optoelectronic, electrochemical, and electromechanical devices with nanoscale dimensions because of the excellent electrical transport and photonic interconnection due to their crystallinity. ZnO QDs and NPs are of great interest because of the three-dimensional confinement of carrier, and phonon leads not only continuous tuning of the optoelectronic properties but also improvement in device performance. As a wide-bandgap semiconductor, ZnO has been reported as an alternative for DSCs because ZnO offers a large direct bandgap which is similar to TiO_2 and even higher electron mobility. ZnO also can be tailored to various nanostructures that provides a promising means for improving the performance of the photoelectrode in DSCs. It is of great urgency to effectively design and control the process window that enables the seamless integration not only the mass production but reproducibility of ZnO nanostructures for the near future.

Author details

Hsin-Ming Cheng* and Shun-Wei Liu*
Department of Electronic Engineering and Organic Electronics Research Center, Ming Chi University of Technology, New Taipei City, Taiwan

*Address all correspondence to: smcheng@mail.mcut.edu.tw
and swliu@mail.mcut.edu.tw

References

[1] Kaidashev EM, Lorenz M, von Wenckstern H, Rahm A, Semmelhack HC, Han KH, et al. High electron mobility of epitaxial ZnO thin films on c-plane sapphire grown by multistep pulsed-laser deposition. Applied Physics Letters. 2003;**82**:3901

[2] Law M, Greene LE, Johnson JC, Saykally R, Yang P. Nanowire dye-sensitized solar cell. Nature Materials. 2005;**4**:455

[3] Baxter JB, Aydil ES. Nanowire-based dye-sensitized solar cells. Applied Physics Letters. 2005;**86**:053114

[4] Cheng HM, Chiu WH, Lee CH, Tsai SY, Hsieh WF. Formation of Branched ZnO Nanowires from Solvothermal Method and Dye-Sensitized Solar Cells Applications. The Journal of Physical Chemistry C. 2008;**112**(42):16359

[5] Wang ZL, Kong XY, Zuo JM. Induced Growth of Asymmetric Nanocantilever Arrays on Polar Surfaces. Physical Review Letters. 2003;**91**:185502

[6] Yan H, He R, Johnson J, Law M, Saykally R, Yang P. Dendritic Nanowire Ultraviolet Laser Array. The Journal of the American Chemical Society. 2003;**125**:4728

[7] Cheng HM, Hsu HC, Chen SL, Wu WT, Kao CC, Lin LJ, et al. Efficient UV photoluminescence from monodispersed secondary ZnO colloidal spheres synthesized by sol-gel method. Journal of Crystal Growth. 2005;**277**:192

[8] Seelig EW, Tang B, Yamilov A, Cao H, RPH C. Self-assembled 3D photonic crystals from ZnO colloidal spheres. Materials Chemistry and Physics. 2003;**80**:257

[9] Ocaña M, Clemente R, Serna CJ. Uniform colloidal particles in solution: Formation mechanisms. Journal of Advanced Materials. 1995;**7**:212

[10] Cheng HM, Lin KF, Hsu HC, Lin CJ, Lin LJ, Hsieh WF. Enhanced Resonant Raman Scattering and Electron–Phonon Coupling from Self-Assembled Secondary ZnO Nanoparticles. The Journal of Physical Chemistry B. 2005;**109**(39):18385

[11] Murray CB, Kagan CR, Bawendi MG. Self-Organization of CdSe Nanocrystallites into Three-Dimensional Quantum Dot Superlattices. Science. 1995;**270**:1335

[12] Murray CB, Sun SH, Gaschler W, Doyle H, Betley TA, Kagan CR. Colloidal synthesis of nanocrystals and nanocrystal superlattices. The IBM Journal of Research and Development. 2001;**45**(1):47

[13] Parke G, Shindo D, Waseda Y, Sugimoto TJ. Internal Structure Analysis of Monodispersed Pseudocubic Hematite Particles by Electron Microscopy. The Journal of Colloid and Interface Science. 1996;**177**:198

[14] Sugimoto T, Wang Y, Itoh H, Muramatsu A. Systematic control of size, shape and internal structure of monodisperse α-Fe2O3 particles. Colloids and Surfaces A: Physicochemical and Engineering Aspects. 1998;**134**:265

[15] Sugimoto T, Sakata K, Muramatsu A. Formation Mechanism of Monodisperse Pseudocubic α-Fe_2O_3 Particles from Condensed Ferric Hydroxide Gel. The Journal of Colloid and Interface Science. 1993;**159**:372

[16] Sugimoto T, Khan MM, Muramatsu A. Formation mechanism of monodisperse peanut-type α-Fe_2O_3 particles from condensed ferric

hydroxide gel. Colloids and Surfaces A: Physicochemical and Engineering Aspects. 1993;**79**:233

[17] Sugimoto T, Muramatsu A, Sakata K, Shindo DJ. Characterization of hematite particles of different shapes. The Journal of Colloid and Interface Science. 1993;**158**:420

[18] Exarhos GJ, Sharma SK. Influence of processing variables on the structure and properties of ZnO films. Thin Solid Films. 1995;**270**:27

[19] Xu XL, Lau SP, Chen JS, Che GY, Tay BKJ. Polycrystalline ZnO thin films on Si (100) deposited by filtered cathodic vacuum arc. Journal of Crystal Growth. 2001;**223**:201

[20] Ashkenov N, Mbenkum BN, Bundesmann C, Riede V, Lorenz M, Spemann D, et al. Infrared dielectric functions and phonon modes of high-quality ZnO films. Journal of Applied Physics. 2003;**93**:126

[21] Damen TC, Porto SPS, Tell B. Raman Effect in Zinc Oxide. Physics Review. 1966;**142**:570

[22] Scott JF. UV Resonant Raman Scattering in ZnO. Physical Review B. 1970;**2**:1209

[23] Zhang XT, Liu YC, Zhi ZZ, Zhang JY, Lu YM, Shen DZ, et al. Resonant Raman scattering and photoluminescence from high-quality nanocrystalline ZnO thin films prepared by thermal oxidation of ZnS thin films. Journal of Physics D. 2001;**34**(24):3430

[24] Ursaki VV, Tiginyanu IM, Zalamai VV, Masalov VM, Samarov EN, Emelchenko GA, et al. Photoluminescence and resonant Raman scattering from ZnO-opal structures. Journal of Applied Physics. 2004;**96**:1001

[25] Ng HT, Chen B, Li J, Han J, Meyyappan M, Wu J, et al. Optical properties of single-crystalline ZnO nanowires on *m*-sapphire. Applied Physics Letters. 2003;**82**:2023

[26] Cheng HM, Hsu HC, Tseng YK, Lin LJ, Hsieh WF. Raman Scattering and Efficient UV Photoluminescence from Well-Aligned ZnO Nanowires Epitaxially Grown on GaN Buffer Layer. The Journal of Physical Chemistry. B. 2005;**109**:8749

[27] Alivisatos AP, Harris TD, Carroll PJ, Steigerwald ML, Brus LE. Electron–vibration coupling in semiconductor clusters studied by resonance Raman spectroscopy. The Journal of Chemical Physics. 1989;**90**:3463

[28] Klein MC, Hache F, Ricard D, Flyzanis C. Size dependence of electron-phonon coupling in semiconductor nanospheres: The case of CdSe. Physical Review B. 1990;**42**:11123

[29] Callender RH, Sussman SS, Selders M, Chang RK. Dispersion of Raman Cross Section in CdS and ZnO over a Wide Energy Range. Physical Review B. 1973;**7**:3788

[30] Alim KA, Fonoberov VA, Balandin AA. Origin of the optical phonon frequency shifts in ZnO quantum dots. Applied Physics Letters. 2005;**86**:053103

[31] Alim KA, Fonoberov VA, Shamsa M, Balandin AA. Micro-Raman investigation of optical phonons in ZnO nanocrystals. Journal of Applied Physics. 2005;**97**:124313

[32] Bartolo BD, Powell R. Phonons and Resonances in Solids. New York: Wiley; 1976. Chap. 10

[33] Böer KW. Survey of Semiconductor Physics. New York: Van Nostrand Reinhold; 1990. Chap. 20

[34] Scamarcio G, Spagnolo V, Ventruti G, Lugara M, Righini GC. Size

dependence of electron—LO-phonon coupling in semiconductor nanocrystals. Physical Review B. 1996;**53**:R10489

[35] Kaminow IP, Johnston WD. Quantitative Determination of Sources of the Electro-Optic Effect in $LiNbO_3$ and $LiTaO_3$. Physics Review. 1967;**160**:519

[36] Loudon R. The Raman effect in crystals. Advances in Physics. 1964;**13**(52):423

[37] Makino T, Tamura K, Chia CH, Segawa Y, Kawasaki M, Ohtomo A, et al. Size dependence of exciton–longitudinal-optical-phonon coupling in ZnO/Mg0.27Zn0.73O quantum wells. Physical Review B. 2002;**66**:233305

[38] Wang RP, Xu G, Jin P. Size dependence of electron-phonon coupling in ZnO nanowires. Physical Review B. 2004;**69**:113303

[39] Wu JJ, Wang DKP. Fabrication and Impedance Analysis of n-ZnO Nanorod/p-Si Heterojunctions to Investigate Carrier Concentrations in Zn/O Source- Ratio-Tuned ZnO Nanorod Arrays. Advanced Materials. 2015;**2007**:19

[40] Wu JJ, Chen GR, Yang HH, Ku CH, Lai JY. Effects of dye adsorption on the electron transport properties in ZnO-nanowire dye-sensitized solar cells. Applied Physics Letters. 2007;**90**:213109

[41] Wang Q, Ito S, Grätzel M, Fabregat-Santiago F, Mora-Seró I, Bisquert J, et al. Characteristics of High Efficiency Dye-Sensitized Solar Cells. The Journal of Physical Chemistry. B. 2006;**110**:25210

[42] Mora-Seró I, Bisquert J, Fabregat-Santiago F, Garcia-Belmonte G, Zoppi G, Durose K, et al. Implications of the Negative Capacitance Observed at Forward Bias in Nanocomposite and Polycrystalline Solar Cells. Nano Letters. 2006;**6**:640

[43] Adachi M, Sakamoto M, Jiu J, Ogata Y, Isoda S. Determination of Parameters of Electron Transport in Dye-Sensitized Solar Cells Using Electrochemical Impedance Spectroscopy. The Journal of Physical Chemistry. B. 2006;**110**:13872

[44] Pasquier AD, Chen H, Lu Y. Dye sensitized solar cells using well-aligned zinc oxide nanotip arrays. Applied Physics Letters. 2006;**89**:253513

[45] Chou TP, Zhang Q, Cao G. Effects of Dye Loading Conditions on the Energy Conversion Efficiency of ZnO and TiO2 Dye-Sensitized Solar Cells. Journal of Physical Chemistry C. 2007;**111**:18804

[46] Cheng HM, Hsieh WF. High-efficiency metal-free organic-dye-sensitized solar cells with hierarchical ZnO photoelectrode. Energy and Environmental Science. 2010;**3**(4):442

[47] Cao H, Zhao YG, Ho ST, Seelig EW, Wang QH, Chang RPH. Random Laser Action in Semiconductor Powder. Physical Review Letters. 1999;**82**:2278

[48] Chou TP, Zhang Q, Fryxell GE, Cao G. Hierarchically Structured ZnO Film for Dye-Sensitized Solar Cells with Enhanced Energy Conversion Efficiency. Advanced Materials. 2007;**19**:2588

[49] Zhang Q, Chou TP, Russo B, Jenekhe SA, Cao G. Aggregation of ZnO nanocrystallites for high conversion efficiency in dye-sensitized solar cells. Angewandte Chemie International Edition. 2008;**47**:2402

[50] Cheng HM, Hsieh WF. Electron transfer properties of organic dye-sensitized solar cells based on indoline sensitizers with ZnO nanoparticles. Nanotechnology. 2010;**21**(48):485202

[51] Ito S, Miura H, Uchida S, Takata M, Sumioka K, Liska P, et al. High-conversion-efficiency

organic dye-sensitized solar cells with a novel indoline dye. Chemical Communications. 2008:5194

[52] Kuang D, Uchida S, Humphry-Baker R, Zakeeruddin SM, Grätzel M. Organic dye-sensitized ionic liquid based solar cells: remarkable enhancement in performance through molecular design of indoline sensitizers. Angewandte Chemie, International Edition. 2008;**47**:1923

[53] Kroeze JE, Hirata N, Koops S, Nazeeruddin MK, Schmidt-Mende L, Grätzel M, et al. Alkyl Chain Barriers for Kinetic Optimization in Dye-Sensitized Solar Cells. Journal of the American Chemical Society. 2006;**128**:16376

[54] Horiuchi T, Miura H, Uchida S. Highly-efficient metal-free organic dyes for dye-sensitized solar cells. Chemical Communications. 2003:3036

Chapter 5

Nanostructures in Dye-Sensitized and Perovskite Solar Cells

Shoyebmohamad F. Shaikh, Nanasaheb M. Shinde, Damin Lee, Abdullah M. Al-Enizi, Kwang Ho Kim and Rajaram S. Mane

Abstract

Due to increase of attention in energy and environmental concerns, there has been much interest developed in clean and renewable energy technologies. The utilization of green and eco-friendly sunlight through solar cells like photovoltaic cells, photo-electrochemical cells, and dye-sensitize and perovskite solar cells (DSSCs and PSCs) produces energy demand. Due to high electron mobility, suitable band alignment, and high optical transparency, the binary and ternary transition metal oxide materials such as TiO_2, SnO_2, ZnO, WO_3, Bi_2O_3 and $SrTiO_3$, Zn_2SnO_4, $BaSnO_3$, etc. have attracted considerable attention as DSSC and PSC electrode materials. Highly efficient solar cells with sustainable performance under severe mechanical deformations are in great demand in forming wearable power supply devices, essential for space technologies. In this regard, myriads of studies have progressed in developing the said metal oxides by various means of nanostructure forms. The aim of this chapter is to highlight research background, basic concepts, operating parameters, working principles, theoretical aspects, and selection of materials with essential properties for DSSCs and PSCs applications.

Keywords: nanostructures, binary and ternary metal oxides, photovoltaic, dye-sensitized solar cells, perovskite solar cells

1. Introduction

1.1 Background and motivation

The predicted global energy needs, due to increasing concerns of environmental pollution in the twenty-first century, have motivated a great deal of efforts into the reduction of fossil fuel consumption followed exploration of clean, renewable, abundant, and eco-friendly renewable energy source technologies to enrich the quality of lives on this planet. Several renewable energy technologies are being investigated to evaluate their potential to address growing demand. These sources include wind turbines, hydropower, nuclear power plants, wave and tidal power, solar cells, solar thermal, and so on. Among these sources, photovoltaic technology, where sunlight is converted into an electrical energy, the so-called solar energy, has drawn considerable attention as it converts into a unique and potential solution. All renewable energy source technologies confirm a common dream, i.e., to capture one type of energy which later can convert into valuable and strategically important asset, that is, electric energy. As the Sun provides a considerable amount of energy for our planet, the energy it provided is approximately 10,000 times more than global demand (i.e., 31,024 J/year); conversion of its 0.1% that is received by the Earth's surface using solar cells with power conversion efficiency of 10% would fulfill our present needs [1].

1.2 Photovoltaic cell

Photovoltaic device generates electrical power by converting sunlight into electricity in the presence of semiconducting materials using the phenomenon, the so-called photovoltaic effect. At first, French scientist Alexandre Becquerel in 1839 discovered photovoltaic effect [2]. After that more than 100 years later, Reynolds et al. in 1954 developed silicon solar cell that was primarily used in space applications until about the mid-1970s [3]. Recently, various kinds of photovoltaic devices are being developed, including silicon solar cells (Si-SCs), dye-sensitized solar cells (DSSCs), quantum dot-sensitized solar cells (QD-DSSCs), organic photovoltaic cells (OPVs), perovskite solar cells (PSCs), etc. Presently, solar cell devices are being used in customer electronics, small-scale remote residential power systems, communications, and signaling applications.

1.3 Photovoltaic parameters

In addition to series and shunt resistance, the photovoltaic solar cells' performance is mainly characterized by six important parameters, (1) short-circuit current density, (2) open-circuit voltage, (3) maximum power output, (4) fill factor, (5) incident photon-to-current conversion efficiency, and (6) solar energy to power conversion efficiency, which are thoroughly discussed as follows [4].

1.3.1 Short-circuit current density

The short-circuit current density is usually written as J_{sc}, which corresponds to the current that passes through the solar cell of one square centimeter area when the impedance is low and voltage across solar cell is zero. The J_{sc} arises due to generation and collection of light-generated charge carriers. For ideal solar cell, at most moderate resistive loss mechanisms, the J_{sc} and light-generated current are identical. Basically, J_{sc} depends upon the area of the solar cell, number of photons reaching at the junction, exposure of incident light, etc.

1.3.2 Open-circuit voltage

The open-circuit voltage is usually presented as V_{oc}, which is the maximum voltage generated from a solar cell when there is no current. The V_{oc} corresponds to the amount of forward bias on solar cell due to bias of solar cell junction with light-generated current. The power (P) produced by solar cell in Watt can be easily calculated alone by *I-V* curve using the equation, $P = IV$. The voltage and current at maximum power from point are denoted as V_{mp} and I_{mp}, respectively.

1.3.3 Maximum power output

For a given bias voltage, the power output of solar cell is the product of measured cell current and voltage. The J_{sc} and V_{oc} are the maximum current and voltage, respectively, from a solar cell. However, at both of these operating points, the power from solar cell is zero.

1.3.4 Fill factor

The fill factor (*ff*) is defined as the ratio of the maximum power from the solar cell to the product of V_{oc} and J_{sc}. Graphically, *ff* is a measure of "squareness" of solar cell and also an area of the largest rectangle which will fit in the *I-V* curve. Typically, the range of *ff* is from 0.50 to 0.82 or 50 to 82% as the *ff* is also represented in percentage using the following relation:

$$ff(\%) = \frac{P_{mp}}{J_{sc}V_{oc}} = \frac{I_{mp}V_{mp}}{J_{sc}V_{oc}} \times 100 \quad (1)$$

1.3.5 Incident photon-to-current conversion efficiency

The incident monochromatic photon-to-current conversion efficiency (IPCE), sometimes referred to also as the "external quantum efficiency" (EQE), is an important parameter of solar cell device. The IPCE can be under light intensity of 100 mW cm^2 (AM1.5), which is estimated in the following equation:

$$IPCE\,(\%) = \frac{J_{sc}(A)}{P(W)} \times \frac{1240}{\lambda(nm)} \times 100 \quad (2)$$

where photocurrent density (*J*) is generated by monochromatic light with wavelength (λ) and intensity (*P*).

1.3.6 Solar-to-electrical power conversion efficiency

The solar-to-electrical power conversion efficiency (η%) is an essential parameter to confirm the performance of one solar cell under testing. The η is defined as the ratio of energy output from the solar cell to an energy input from the Sun. In addition to reflecting performance of the solar cell itself, η depends on the spectrum and intensity of the incident sunlight and temperature of the solar cell. Therefore, terrestrial solar cells are measured under air mass 1.5 conditions in addition to a temperature of 25°C. The solar cells intended for space use are measured under air mass "0" conditions. It is well known that an overall η (%) of the solar cells can be determined by J_{sc}, V_{oc}, *ff*, and the intensity of the incident light (P_{in}) as follows:

$$\eta(\%) = \frac{P_{out}}{P_{in}} = \frac{J_{sc}V_{oc}ff}{P_{in}} \times 100 \quad (3)$$

1.4 History of dye-sensitized and perovskite solar cells

In 1873, Vogel and Berlin invented dye sensitization technique, but until the 1970s DSSCs' mechanism was unclear. Therefore, compared with silicon-based photovoltaic devices, the performance of these early DSSCs was poor (η = <1%). The major obstacle in poor performance was relatively low adsorption of dye molecules into the metal oxide photoanode surface. Few improvements in efficiency were achieved by coating a thick layer of dye molecules onto the metal oxide photoanode surface [5]. Nevertheless, the power conversion efficiency was limited to ≤2% due to the low-light harvesting and charge collection from the adsorbed dye molecules. In the 1990s, Prof. Michel Gratzel and his team creatively demonstrated a practical photoelectrochemical cell device with a certified power conversion efficiency of 11.9%, presenting excellent market competitiveness and commercial prospect [6]. However, such DSSCs of a liquid-state electrolyte are with highly volatile solvents, which not only affect the long-term stability of the device but also limit its large-scale production. In 1998, Gratzel et al. unveiled a solid-state organic hole-transporting material, i.e., 2,2,7,7-tetrakis (N,N-di-p-methoxy-phenylamine)-9,9-spirobifluorene (spiro-OMeTAD), to replace the conventional liquid-state electrolyte for a solid-state DSSCs [7]. From 1998 to 2011, the power conversion efficiency of solid-state DSSCs increased steadily from 0.74% to 7.2% but still much lower than that obtained by liquid-state electrolyte-based DSSCs [8]. In 2009, Miyasaka and co-workers improved the power conversion efficiency of perovskite solar cell to 3.8% by replacing bromine ($MAPbBr_3$) with iodine ($MAPbI_3$) [9]. In 2011, Park et al. prepared $MAPbI_3$ perovskite as quantum dots with a size of ca. 2–3 nm, resulting in an enhanced power conversion efficiency of 6.5% [10]. In the family of photovoltaic device, PSCs have demonstrated much stronger absorption than the standard N719 dye, but the limitation in perovskite-based solar cells was the rapid degradation of the device performance resulting in the dissolution of perovskites in liquid-state electrolytes. In 2012, Park's group works together with Gratzel's group and introduced perovskite-sensitized solar cell by using solid-state spiro-OMeTAD as a hole-transporting material by replacing liquid-state electrolytes [11], where the solid-state spiro-OMeTAD not only solved the problem of perovskite dissolution but also significantly improved the stability and power conversion efficiency (9.7%). Remarkable progress of perovskite solar cell has been made during 2013–2014 as the power conversion efficiency increased to a certified 16.2% and 20.1% [12, 13]. Sahil et al. prepared fully textured monolithic perovskite/silicon tandem solar cells with 25.2% certified power conversion efficiency. Besides the breakthrough in efficiency, novel designs of device architectures aiming for low-cost and highly stable DSSCs and PSCs have also been developed [14].

1.5 Theoretical aspect

After the invention of nanostructure-based photovoltaic solar cells, a lot of theoretical and experimental works has been carried out to explain its operation principle. The need for unique theoretical considerations of photovoltaic effect arises from the fundamental differences in the operation between DSSCs and PSCs over traditional semiconductor *p-n and p-i-n* junction solar cells [15].

1.5.1 Light absorption

In photovoltaic solar cell, light absorption and charge transport occur in the same material, whereas in the DSSCs, photons get absorbed by dye molecules and charge transport is carried out in photoanode and electrolyte. The high efficiency of DSSCs is accomplished by coating the internal surfaces of porous metal oxide-based photoanode with special dye (N3, N719, and black dye) molecules which are tuned to absorb incoming photons of all wavelengths. The absorption of a photon by dye molecule takes place *via* an excitation between the electronic states of molecule. Similarly, in PSCs, the absorption spectrum demonstrates good light-harvesting capabilities over the visible to near-IR spectrum which is also stable during prolonged light exposure. For sensitized mesoporous metal oxide devices, it is inferred that after light absorption in the perovskite, electrons are transferred to the metal oxide-based electron transfer layer followed by the conducting substrate (i.e., fluorine-tin-oxide/indium-tin-oxide), and holes are transferred to the spiro-OMeTAD and then to the silver/gold contact electrode for driving the applied load. The enhancement in light absorption near the band edge can be carefully engineered through various nanostructures for better photon management to increase the current followed by power conversion efficiency.

1.5.2 Charge separation

As the next step of the conversion of light absorption into electrical current, a complete charge separation must be achieved. The charge separation in photovoltaic solar cell is induced by the electric field across the junction, while no such long-range electric fields are found in the DSSCs and PSCs. The charge separation in the DSSCs is basically an electron transfer process from dye molecule to photoanode and hole transport process from oxidized dye to electrolyte. The electron transfer mechanism depends on an electronic structure of dye molecules and energy level matching between the excited state of dye and conduction band of photoanode, i.e., metal oxide. The lowest unoccupied molecular orbital (LUMO) should be above the conduction band edge of photoanode, and the highest occupied molecular orbital (HOMO) should be below the chemical potential of redox pair of electrolyte, i.e., iodide/triiodide, which is supposed to be an energetic driving force for electron and hole separation. In addition, entropic factors play an important role for energetic charge separation. The large density of delocalized states in the metal oxide nanoparticles compared with dye molecules on the surface facilitates electron injection in its conduction band, which eventually increases the driving force in entropy (approximately 0.1 eV) for charge separation. In PSCs, the charge separation and transportation occur between metal oxide as an electron transfer layer and spiro-OMeTAD as a hole transfer layer interfacial surface. On exposing to the Sun's radiation, photo-excited electrons are injected from perovskite absorber layer into the conduction band of the metal oxide electron transfer layer, and the hole is transported to spiro-OMeTAD hole transfer layer followed by the charge collector for driving efficiently.

1.5.3 Recombination

Figure 1 presents the possible ways of recombination: (1) electron injection from dye-excited state to the conduction band of photoanode, (2) regeneration of dye cation by electron transfer from the redox couple, (3) charge recombination to the cation of dye, (4) recombination to the redox couple, and (5) excited-state decay to the ground state. The photo-injected electrons in the photoanode can have

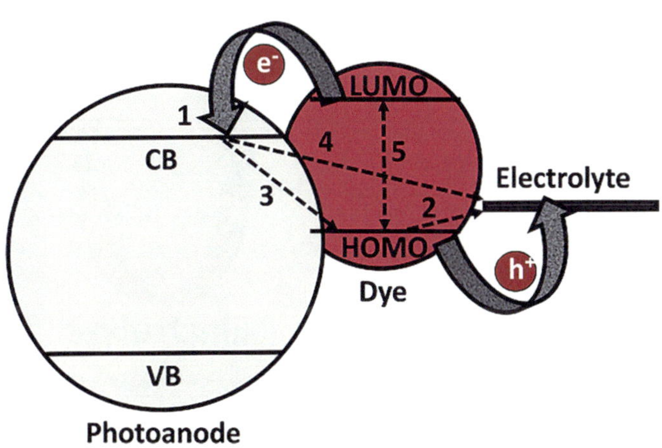

Figure 1.
Plausible electron recombination processes in the DSSCs.

two possible recombination pathways: direct recombination with the oxidized dyes or with the I_3^- in the electrolyte (**Figure 1**). The latter process is dominant and has thoroughly been studied in the literature by many researches.

1.5.4 Air mass

In Astronomy, air mass is direct path length through the earth atmosphere, expressed as a ratio relative to path length vertically upward, i.e., at the zenith. The intensity of solar radiations decreases with distance. Therefore, different air mass standards are being formulated to account for obstruction caused by the Earth's atmosphere.

The air mass standard is denoted by "AM-X" in which X represents air mass coefficient:

$$X = \frac{1}{\cos\theta} \tag{4}$$

where θ is called solar zenith angle. This is defined as the angle between normal of a given point on the earth and light path coming to that point from the Sun. The air mass standards are categorized into three different kinds: briefly, the AM 0 spectrum, for the solar radiation outside the atmosphere; AM 1 for flux of solar energy normal to the Earth; and AM 1.5 represents the solar energy flux impinging at the Earth's surface with 48.2° zenith angle.

2. Working principal of the DSSCs

The working principal of the DSSCs is demonstrated in **Figure 2**. The wide bandgap nanocrystalline TiO_2 (photoanode) semiconductor film is needed to be deposited on the conducting substrate (FTO) either by direct deposition or by doctor-blade method to provide the necessary large surface area to adsorb sensitizers (dye molecules). Upon absorption of photons, dye molecules are excited from

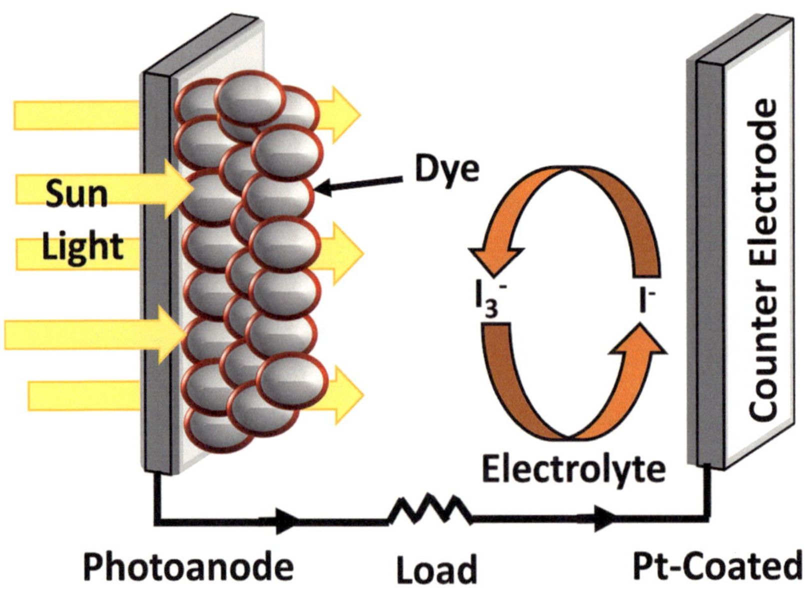

Figure 2.
The operating principle of the DSSCs.

the HOMO to the LUMO. Once an electron is injected into the conduction band of the TiO_2 photoanode, the dye molecule (photosensitizer) is oxidized. The injected electron is then transferred to TiO_2 nanostructured photoanode through hopping kinetics which is finally extracted at load, where the work done is delivered as an electrical energy. The electrolytes containing I^-/I_3^- redox ions are used as an electron mediator between the TiO_2 photoanode and the platinum-coated counter electrode. Therefore, the oxidized dye molecules (photosensitizer) are regenerated by receiving electrons from the I^- ion redox mediator that get oxidized to I_3^- (triiodide ions) [16]. Regarding the working mechanism of perovskite solar cell, various research groups reported differently; thereby, it is yet to be unprecedented.

3. Nanostructures of inorganic materials

A major feature that discriminates various types of nanostructures is their dimensionality (**Figure 3a–d**). The word "nano" stems from the Greek word "nanos," which means dwarf. This word "nano" has been assigned to indicate the number 10^{-9}, i.e., 1 billionth of any unit [17]. It is believed that the size of particle has inverse relation with the surface area and reactivity thereby, nanoparticles reveal superior practical potential over micron-sized one.

3.1 Zero-dimensional nanostructures

The significant progress has been made in the field of zero-dimensional nanostructures. A rich variety of physical and chemical methods have been proposed to synthesize zero-dimensional nanostructures. Recently, zero-dimensional nanostructures such as uniform particle arrays (quantum dots), heterogeneous particle arrays, core-shell quantum dots, onions, and hollow spheres are being obtained by several research groups [18].

3.2 One-dimensional nanostructures

In the last decade, one-dimensional nanostructures have stimulated an increasing attention due to their importance in research and a wide range of potential

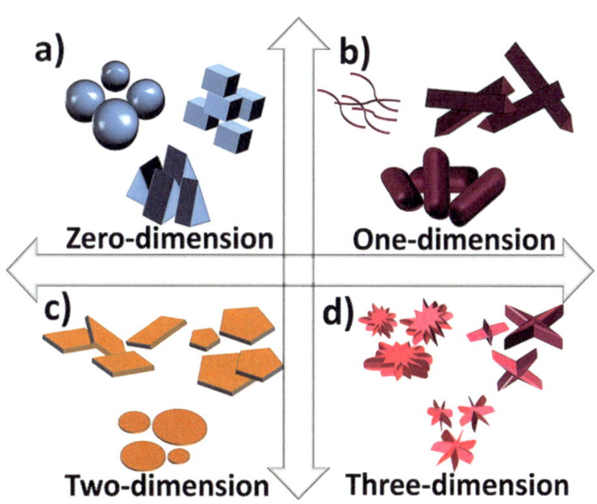

Figure 3.
(a–d) Nanostructures of different dimensions.

applications. It is generally accepted that one-dimensional nanostructures are ideal for exploring a large number of novel phenomena at the nanoscale level and corroborating the size and dimensionality dependence of functional properties. In **Figure 3b**, one-dimensional nanostructures are endowed with typical spherical, pseudo-spherical, dodecahedral, tetrahedral, octahedral, cubic, and corresponding hollow shapes. One-dimensional nanostructures/morphologies also include nanotubes, nano-needles, nano-rods or nano-wires, nano-shuttles, nano-capsules, hollow structures, etc. [19].

3.3 Two-dimensional nanostructures

Two-dimensional nanostructures have two dimensions outside of the nanometric size range. In recent years, a synthesis of two-dimensional nanostructures with certain geometries exhibits unique shape-dependent characteristics and their subsequent utilization as building blocks for the key components of nano-devices. In **Figure 3c**, two-dimensional nanostructures, such as junctions (continuous islands), branched structures, nano-prisms, nano-plates, nano-sheets, nano-walls, nanodisks, etc., are confirmed in the literature. Round disks, hexagonal/triangular/quadrangular plates or sheets, belts, mesoporous hollow nanospheres, hollow rings, etc. are also forms of two-dimensional nanostructures [20].

3.4 Three-dimensional nanostructures

Owing to the large specific surface area and other superior properties over the bulk counterparts arising from the quantum size effect, three-dimensional nanostructures have attracted considerable research interest, and many three-dimensional nanostructures have been synthesized in the past decade (**Figure 3d**). It is well-known that the surface area, shape, size, dimensionality, and morphologies of the nanostructures are key factors to obtain better performance of the device when they are envisaged. As these materials offer higher surface area, they can supply enough absorption sites for all involved molecules in a small space. On the other hand, such materials with higher porosity can lead to a better transportation of dye molecules. A typical three-dimensional nanostructured such as nanocoils, nanocones, nanoflowers, and nanoballs (dendritic structures) are on a great demand [21].

4. Material selection

4.1 Binary transition metal oxides

The binary transition metal oxide materials play an important role in the DSSCs and PSCs (**Figure 4a**). Titanium dioxide (TiO_2) deserves special attention since its cheap, non-toxicity, abundant, biocompatible, facile preparation with diverse morphologies, stability in both acidic and alkaline media features. The TiO_2 exists naturally in three crystalline polymorphs, namely, rutile (E_g = 3.05 eV), anatase (E_g = 3.23 eV), and brookite (E_g = 3.26 eV), and the uniqueness of each lattice structure leads to multifaceted physicochemical and optoelectronic properties [22]. These interesting properties reveal different functionalities, thus influencing their performances in various applications. For instance, rutile phase of TiO_2 exhibits a high refractive index and UV absorptivity and is thus capable of being applied in optical communication devices (isolators, modulators, switches, etc.). Meanwhile, anatase is largely preferred in photovoltaics and photocatalysis because of its superior electron mobility, surface chemistry, potentially higher conduction band edge energy, and catalytic activity compared with the other two phases [23]. The problems associated

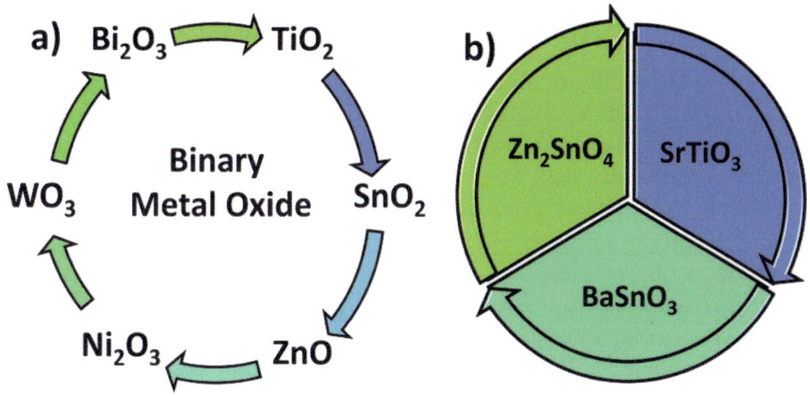

Figure 4.
(a and b) Materials used for the DSSCs and PSCs.

with TiO_2 metal oxide such as high surface state and fast electron recombination rate contribute to adverse effects on the electron mobility and charge transport kinetics. Zinc oxide (ZnO), an important II–VI semiconductor with a wide bandgap of 3.37 eV, similar to TiO_2, has high electron mobility (~155–1000 $cm^2\ V^{-1}\ s^{-1}$) and a large exciton binding energy of 60 meV [24]. Moreover, the electron injection efficiency of ZnO is almost equivalent to that of TiO_2. The electron lifetime of ZnO is significantly higher, and the recombination rate is lower than TiO_2. Nevertheless, ZnO-based DSSCs also have suffered from chemical instability in acidic electrolytes and thereby showed slow electron injection kinetics from dye to ZnO photoanode due to the formation of an insulating surface agglomeration layer. Alternatively, tin oxide (SnO_2) can be a good choice as it is an *n*-type and wide bandgap semiconductor with excellent optical and electrical properties compared to TiO_2. The electron mobility in SnO_2 (~100–250 $cm^2\ V^{-1}\ s^{-1}$) is two orders of magnitude higher than TiO_2 (~0.1–1.0 $cm^2\ V^{-1}\ s^{-1}$), suggesting a faster diffusion transport of the photo-induced electrons. Secondly, SnO_2 has a higher bandgap (3.4 eV) than anatase TiO_2 (3.2 eV), which creates fewer oxidative holes in the valence band (fewer oxidative holes facilitate long-term stability and higher stability under long-term UV irradiation) [25]. The Niobium oxide (Nb_2O_5) is another wide bandgap semiconductor with 3.49 eV bandgap energy, which is nearly 0.29 eV larger than that of TiO_2 (anatase). Because of its larger bandgap and higher conduction band edge compared with anatase TiO_2 it is used to achieve relatively higher V_{oc} than anatase TiO_2 [26]. Recently, Tungsten oxide (WO_3) has attracted immense attention due to its 2.8 eV bandgap energy, which would theoretically utilize ~12% of incident solar light of the visible region. In comparison with TiO_2, WO_3 possesses a higher mobility and has its conduction band edge at a more-positive location (~0.5 V). Therefore, it is speculated that the V_{oc} in WO_3 nanostructured electrode is limited due to the lower difference between its conduction band and redox potential of electrolytes [27]. Bismuth oxide (Bi_2O_3) has several advantages due to its unique electrical, optical, and mechanical properties. It exists in four crystal phases, i.e., monoclinic α-Bi_2O_3, tetragonal β-Bi_2O_3, cubic γ-Bi_2O_3, and cubic δ-Bi_2O_3. The α-Bi_2O_3 phase is most stable at low temperatures up to 730°C, while δ-Bi_2O_3 phase is stable when the temperature is above 1000 K. The β-Bi_2O_3 and γ-Bi_2O_3 phases are high-temperature metastable phases. The Bi_2O_3 also exhibits a high refractive index, dielectric permittivity, high oxygen ion conductivity, and remarkable photoconductivity and photoluminescence [28]. Its bandgap energy of 2.5–3.1 eV mostly depends on the crystal phase type. The narrow bandgap of Bi_2O_3 makes it suitable for a large range of applications including optical coatings, photovoltaics, microwave-integrated circuits, superconductor, etc.

4.2 Ternary transition metal oxides

Besides the simple binary metal oxide systems, ternary metal oxide systems such as Strontium titanate ($SrTiO_3$), Zinc Stannate (Zn_2SnO_4), and Barium Stannate ($BaSnO_3$) have also been considered as photoanode materials in the DSSCs and PSCs (**Figure 4b**). The $SrTiO_3$ is a semiconductor with bandgap similar of 3.2 eV. However, its conduction band is relatively at higher position than that of TiO_2, which results in a higher V_{oc} [29]. A high dielectric constant makes $SrTiO_3$ as electrically mesoporous even with a large particle size of ~80 nm [30]. In addition, Zn_2SnO_4 is particularly interesting because of its physical and electrical properties. The 3.6 eV bandgap and 10–15 $cm^2\ V^{-1}\ s^{-1}$ electron mobility of Zn_2SnO_4 have made it stable against UV light with high, electrical conductivity, and low visible absorption over TiO_2 [31]. The ternary $BaSnO_3$ is an n-type semiconductor with a wide bandgap of 3.1 eV, and its band structure and electrical properties can be controlled easily by atomic substitution or doping into the Ba or Sn site for better performance when used in DSSCs' application [32]. In this sense, as the electrode materials in DSSCs, the ternary oxides are better than the binary.

5. Conclusions

The DSSC and PSC solar cells have attracted scientific and technological importance as an alternative to conventional Si-based solar cells. A market feasibility of the solar cells will be a part of the manufacturing cost, durability, fabrication time, chemical stability, mechanical robustness, and power conversion efficiency. The design strategy, preparation method, and surface chemistry of transition metal oxides with excellent electrical and optical properties will also have an impact. An era of nanotechnology has opened a door to tailing transition metal oxide materials for DSSCs' and PSCs' applications. In this chapter, we briefly have discussed four basic topics about the DSSCs and PSCs. Initially, background, motivation, and present needs of DSSCs and PSCs are covered. The required photovoltaic parameters including short-circuit current density, open-circuit voltage, fill factor, and incident photon-to-current conversion efficiency to develop good DSSCs and PSCs are emphasized in brief. The historical background has been presented to get an idea regarding the new investigations taking place to replace dye molecules through perovskite absorber layer. Information on the theoretical and practical details has also been provided to obtain DSSCs and PSCs with high solar-to-electricity power conversion efficiencies. Working principle of the DSSCs is explored by considering electron and hole pair generation, charge transportation, and charge separation and recombination etc. In transition metal oxide in binary and ternary forms like zero, one, two, and three dimensions such as nanoparticles, nanotubes, nanodisks, and nanoflowers, their implication is proposed as good candidates in developing a smart and wearable DSSCs and PSCs in the future.

Acknowledgements

Author SFS would like to thank University Grants Commission, New Delhi, for awarding Dr. D. S. Kothari Postdoctoral Fellowship scheme (F.4-2/2006 (BSR)/CH/16-17/0015) and author KHK would like to thank Global Frontier Program through the Global Frontier Hybrid Interface Materials (GFHIM) of the National Research Foundation of Korea (NRF) funded by the Ministry of Science, ICT & Future Planning (2013M3A6B1078874).

Conflict of interest

The authors declare no competing interests.

Notes/thanks/other declarations

Authors SFS and RSM would like to thank Mr. S. S. Sangale and Mr. Y. T. Nakate for arranging the references and designing the cartoon images.

Acronyms and abbreviations

AM	Air mass
DSSCs	Dye-sensitized solar cells
e^-	Electron
E_g	Bandgap
EQE	External quantum efficiency
ff	Fill factor
FTO	Fluorine-tin-oxide
h^+	Hole
HOMO	Highest occupied molecular orbital
I-V	Current-voltage
IPCE	Incident photon-to-current conversion efficiency
I_{mp}	Current at maximum power
J_{sc}	Short-circuit current density
J	Photocurrent density
LUMO	Lowest unoccupied molecular orbital
η	Solar-to-electrical power conversion efficiency
OPVs	Organic Photovoltaic Cells
PSCs	Perovskite solar cells
P	Power
QD-DSSCs	Quantum dot-sensitized solar cells
Si-SCs	Silicon solar cells
V_{oc}	Open-circuit voltage
VB	Valence band
V_{mp}	Voltage at maximum power
λ	Wavelength

Author details

Shoyebmohamad F. Shaikh[1,2], Nanasaheb M. Shinde[3], Damin Lee[3], Abdullah M. Al-Enizi[2], Kwang Ho Kim[3] and Rajaram S. Mane[1*]

1 School of Physical Sciences, Swami Ramanand Teerth Marathwada University, Nanded, MH, India

2 Chemistry Department, College of Sciences, King Saud University, Riyadh, Saudi Arabia

3 National Core Research Center, Pusan National University, Busan, Republic of Korea

*Address all correspondence to: rajarammane70@srtmun.ac.in
and kwhokim@pusan.ac.kr

References

[1] Gratzel M. Photoelectrochemical cell. Nature. 2001;**414**:338-344. DOI: 10.1038/35104607

[2] Shah A, Torres P, Tscharner R, Wyrsch N, Keppner H. Photovoltaic technology: The case for thin-film solar cells. Science. 1999;**285**:692-698. DOI: 10.1126/science.285.5428.692

[3] Reynolds DC, Leies G, Antes LL, Marburger RE. Photovoltaic effect in cadmium sulfide. Physics Review. 1954;**96**:533-534. DOI: https://doi.org/10.1103/PhysRev.96.533

[4] Hua Y. Modification of TiO_2 photoanodes for dye-sensitized solar cells [thesis]. Griffith School of Environment Science, Environment, Engineering and Technology: Griffith University; 2010

[5] Flynn BW, Owen AE, Mavor J. Dye-sensitized of the photoconductivity of SiO_2 films in M-dye-SiO_2-M structure. Journal of Physics C: Solid State Physics. 1977;**10**:4051-4059. DOI: 10.1088/0022-3719/10/20/013

[6] Oregan B, Gratzel M. A low-cost, high efficiency solar cell based on dye-sensitized colloidal TiO_2 films. Nature. 1991;**353**:737-739. DOI: 10.1038/353737a0

[7] Bach U, Lupo D, Comte P, Moser JE, Weissortel F, Salbeck J, et al. Solid-state dye-sensitized mesoporous TiO_2 solar cells with high photon-to-electron conversion efficiencies. Nature. 1998;**395**:583-585. DOI: 10.1038/26936

[8] Burschka J, Dualeh A, Kessler F, Baranoff E, Cevey-Ha NL, Yi C, et al. Tris (2-(1 H-pyrazol-1-yl) pyridine) cobalt (III) as p-type dopant for organic semiconductors and its application in highly efficient solid-state dye-sensitized solar cells. Journal of the American Chemical Society. 2011;**133**:18042-18045. DOI: 10.1021/ja207367t

[9] Kojima A, Teshima K, Shirai Y, Miyasaka T. Organometal halide perovskites as visible-light sensitizers for photovoltaic cells. Journal of the American Chemical Society. 2009;**131**:6050-6051. DOI: 10.1021/ja809598r

[10] Im JH, Lee CR, Lee JW, Park SW, Park NG. 6.5% efficient perovskite quantum-dot-sensitized solar cell. Nanoscale. 2011;**3**:4088-4093. DOI: 10.1039/C1NR10867K

[11] Kim H, Lee C, Im J, Lee K, Moehl T, Marchioro A, et al. Lead iodide perovskite sensitized all-solid-state submicron thin film mesoscopic solar cell with efficiency exceeding 9%. Scientific Reports. 2012;**2**:591. DOI: 10.1038/srep00591

[12] Jeon NJ, Noh JH, Kim YC, Yang WS, Ryu S, Seok SI. Solvent engineering for high-performance inorganic-organic hybrid perovskite solar cells. Nature Materials. 2014;**13**:897-903. DOI: 10.1038/nmat4014

[13] Yang W, Noh J, Jeon N, Kim Y, Ryu S, Seo J, et al. High-performance photovoltaic perovskite layers fabricated through intramolecular exchange. Science. 2015;**348**:1234-1237. DOI: 10.1126/science.aaa9272

[14] Sahli F, Werner J, Kamino A, Bräuninger M, Monnard R, Paviet-Salomon B, et al. Fully textured monolithic perovskite/silicon tandem solar cells with 25.2% power conversion efficiency. Nature Materials. 2018;**1**:820-826. DOI: 10.1038/s41563-018-0115-4

[15] Shaikh SF. Chemical synthesis of metal oxides for DSSCS application: Effect of interfacial surface modification on power conversion

efficiency [thesis]. Clean Energy and Chemical Engineering, University of Science and Technology; 2015

[16] Nazeeruddin Md K, Baranoff E, Gratzel M. Dye-sensitized solar cells: A brief overview. Solar Energy. 2011;**85**:1172-1178. DOI: 10.1016/j.solener.2011.01.018

[17] Tiwari JN, Tiwari RN, Kim KS. Zero-dimensional, one-dimensional, two-dimensional and three-dimensional nanostructured materials for advanced electrochemical energy devices. Progress in Materials Science. 2012;**57**:724-803. DOI: 10.1016/j.pmatsci.2011.08.003

[18] Kim YT, Han JH, Hong BH, Kwon YU. Electrochemical synthesis of CdSe quantum-dot arrays on a graphene basal plane using mesoporous silica thin-film templates. Advanced Materials. 2010;**22**(4):515-518. DOI: https://doi.org/10.1002/adma.200902736

[19] Kim WY, Choi YC, Kim KS. Understanding structures and electronic/spintronic properties of single molecules, nanowires, nanotubes, and nanoribbons towards the design of nano devices. Journal of Materials Chemistry. 2008;**18**(38):4510-4521. DOI: 10.1039/B804359K

[20] Jung SH, Oh E, Lee KH, Yang Y, Park CG, Park W, et al. Sonochemical preparation of shape-selective ZnO nanostructures. Crystal Growth and Design. 2007;**8**(1):265-269. DOI: 10.1021/cg070296l

[21] Ahn JH, Kim HS, Lee KJ, Jeon S, Kang SJ, Sun Y, et al. Heterogeneous three-dimensional electronics by use of printed semiconductor nanomaterials. Science. 2006;**314**(5806):1754-1757. DOI: 10.1126/science.1132394

[22] Luo H, Takata T, Lee Y, Zhao J, Domen K, Yan Y. Photocatalytic activity enhancing for titanium dioxide by co-doping with bromine and chlorine. Chemistry of Materials. 2004;**16**:846-849. DOI: 10.1021/cm035090w

[23] Shaikh SF, Mane RS, Min BK, Hwang YJ, Joo OS. D-sorbitol-induced phase control of TiO_2 nanoparticles and its application for dye-sensitized solar cells. Scientific Reports. 2016;**6**:20103. DOI: 10.1038/srep20103

[24] Zheng YZ, Tao X, Hou Q, Wang DT, Zhou WL, Chen JF. Iodine-doped ZnO nanocrystalline aggregates for improved dye-sensitized solar cells. Chemistry of Materials. 2011;**23**:3-5. DOI: 10.1021/cm101525p

[25] Shaikh SF, Mane RS, Joo OS. Mass scale sugar-mediated green synthesis and DSSCs application of tin oxide nanostructured photoanode: Effect of zinc sulphide layering on charge collection efficiency. Electrochimica Acta. 2014;**147**:408-417. DOI: 10.1016/j.electacta.2014.08.146

[26] Lenzmann F, Krueger J, Burnside S, Brooks K, Gratzel M, Gal D, et al. Surface photovoltage spectroscopy of dye-sensitized solar cells with TiO_2, Nb_2O_5, and $SrTiO_3$ nanocrystalline photoanodes: Indication for electron injection from higher excited dye states. The Journal of Physical Chemistry. B. 2001;**10**:6347-6352. DOI: 10.1021/jp010380q

[27] Hara K, Zhao Z, Cui Y, Miyauchi M, Miyashita M, Mori S. Nanocrystalline electrodes based on nanoporous-walled WO_3 nanotubes for organic-dye sensitized solar cells. Langmuir. 2011;**27**:12730-12736. DOI: 10.1021/la201639f

[28] Shaikh SF, Rahman G, Mane RS, Joo OS. Bismuth oxide nanoplates-based efficient DSSCs: Influence of ZnO surface passivation layer. Electrochimica Acta. 2013;**111**:593-600. DOI: 10.1016/j.electacta.2013.08.066

[29] Sun Q, Hong Y, Zang T, Liu Q, Yu L, Dong L. The application of heterostructured $SrTiO_3$-TiO_2 nanotube arrays in dye-sensitized solar cells. Journal of the Electrochemical Society. 2018;**165**(4):H3069-H3075. DOI: 10.1149/2.0101804jes

[30] Hod I, Shalom M, Tachan Z, Rühle S, Zaban A. $SrTiO_3$ recombination-inhibiting barrier layer for type II dye-sensitized solar cells. Journal of Physical Chemistry C. 2010;**114**:10015-10018. DOI: 10.1021/jp101097j

[31] Chen J, Lu L, Wang W. Zn_2SnO_4 nanowires as photoanode for dye sensitized solar cells and the improvement on open-circuit voltage. Journal of Physical Chemistry C. 2012;**116**:10841-10847. DOI: 10.1021/jp301770n

[32] Shin SS, Kim JS, Suk JH, Lee KD, Kim DW, Park JH, et al. Improved quantum efficiency of highly efficient perovskite $BaSnO_3$-based dye-sensitized solar cells. ACS Nano. 2013;**7**:1027-1035. DOI: 10.1021/nn305341x

Chapter 6

Structural Engineering on Pt-Free Electrocatalysts for Dye-Sensitized Solar Cells

Yi-June Huang, Han-Ting Chen, Shiuan-Bai Ann, Chun-Ting Li and Chuan-Pei Lee

Abstract

In recent decades, plenty of nanomaterials have been investigated as electrocatalysts for the replacement of the expensive platinum (Pt) counter electrode in dye-sensitized solar cells (DSSCs). The key function of the electrocatalyst is to reduce tri-iodide ions to iodide ions at the electrolyte/counter electrode interface. The performance of the electrocatalyst is usually determined by two key factors, i.e., the intrinsic heterogeneous rate constant and the effective electrocatalytic surface area of the electrocatalyst. The intrinsic heterogeneous rate constant of the electrocatalyst varies by different types of materials, which can be roughly divided into five groups: non-Pt metals, carbons, conducting polymers, transition metal compounds, and their composites. The effective electrocatalytic surface area is determined by the nanostructure of the electrocatalyst. In this chapter, the nanostructural design and engineering on different types of Pt-free electrocatalysts will be systematically introduced. Also, the relationship between various nanostructures of electrocatalysts and the pertinent physical/electrochemical properties will be discussed.

Keywords: counter electrode, dye-sensitized solar cells, electrocatalyst, nanostructure

1. Introduction

1.1 Dye-sensitized solar cells (DSSCs)

Fossil fuel, as a limiting energy source, may be run out in the upcoming centuries. However, the consumption of energy increases every year [1, 2]. As a result, finding and developing renewable energy sources is an urgent problem. Due to the unlimitedness of renewable energy resources, they are candidates to be reliable replacement for sustainable usage in the future. Among them, the Sun has been considered as one of the most promising renewable energy sources. It provides about 120,000 terawatts to the earth, which equals thousand times of the current energy consumption rate. The solar cells can utilize the sunshine and transform to electricity [3–5]. Generally, solar cells can be classified to four generations: the first generation is silicon-based solar cells; the second generation is CIGS (CuInGaSe), CZTS (CuZnTiSe), and CdTe solar cells; the third generation is organic photovoltaics (OPVs) and dye-sensitized solar cells (DSSCs); and the fourth generation is perovskite solar cells (PSCs). The first and second generations have been widely explored for decades, and they are the most common solar cells at present. However, they are fabricated through expensive, toxic, energy-intensive, high-temperature, and high-vacuum processes. Therefore, DSSCs are very competitive to the first and second generations of solar cells due to numerous advantages including easy fabrication, low cost (**Figure 1**) [3], and high performance at dim-light condition. Moreover, DSSCs can be used in indoor ambient applications [5–9].

A DSSC is composed of a photoanode, electrolyte, and counter electrode (CE), as shown in **Figure 2**. When a photoanode is excited by the sun or photon, it will release the electron to the external circuit. At the same time, the iodide/triiodide (I^-/I_3^-) redox couple will relax the photoanode to its ground state. Then the CE will reduce the redox couple to regenerate the DSSC. Among them, the CE plays an important role to determine the DSSC performance [10]. At the CE/electrolyte interface, the electrochemical mechanism goes through I_3^- decomposition (Eq. (1)) → adsorption (Eq. (2)) → catalytic reduction reaction (Eq. (3)) → desorption (Eq. (4)), and the overall reaction shows as Eq. (5) [11]. Among these reaction steps, Eq. (3) is found to be the slowest step, which means the rate-determining step to decide the DSSC performance.

$$I_3^- \leftrightarrow I_2 + I^- \tag{1}$$

$$I_2 + 2CE \leftrightarrow I\,(CE) + I\,(CE) \tag{2}$$

$$I\,(CE) + e^- \leftrightarrow I^-\,(CE) \tag{3}$$

$$I^-\,(CE) \leftrightarrow I^- + CE \tag{4}$$

$$I_3^- + 2e^- \leftrightarrow 3I^- \tag{5}$$

There are two ways to enhance the electrocatalytic reduction reaction. One is to increase the heterogeneous rate constant, relating to the intrinsic electrocatalytic ability of the electrocatalyst. The other is to engineer the structure of the electrocatalyst for I_3^- reduction with regard to the charge transfer route and the surface area. To replace a traditional platinum (Pt) electrocatalyst, where Pt is a rare and expensive

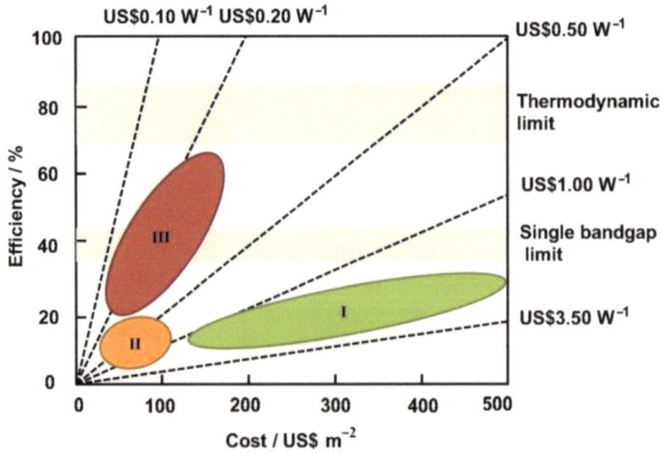

Figure 1.
Efficiency and cost projection for first- (I), second- (II), and third-generation (III) photovoltaic technology [3].

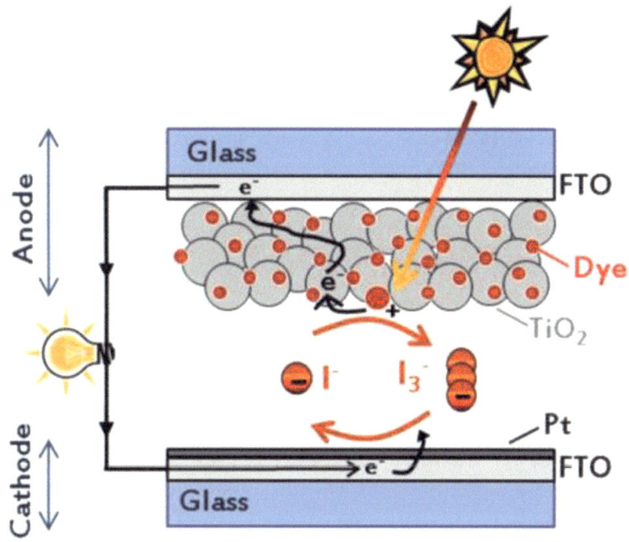

Figure 2.
The structure and mechanism of DSSCs [2].

element, several types of materials, such as carbon materials [5, 12–28], conductive polymers [29–48], and transition metal compounds [37, 45, 49–79] have been extensively explored to elevate the cell efficiency (η) and decrease the cost of the CEs.

To date, there have been a very limited number of non-Pt nanomaterials that could have a comparable intrinsic heterogeneous rate constant to that of Pt. However, the specific structural designs of nanomaterials would largely increase the effective electrocatalytic surface area so as to provide better overall electrocatalytic ability than Pt. Moreover, the nanostructured electrocatalysts could have appropriate interfacial affinity, good electrochemical stability, or specific self-assembly natures; these properties may influence the DSSC performance as well. A typical nanostructured material can be defined if any dimension of the material is lower than 100 nm. The nanostructured material can be classified into three groups: zero-dimensional (0D, e.g., nanoparticle, nanocube, etc.), one-dimensional (1D, e.g., nanorod, nanotube, nanoneedle, etc.), and two-dimensional (2D, e.g., nanosheet, nanopental, etc.) structures, as shown in **Figure 3**. Generally, 0D structure is

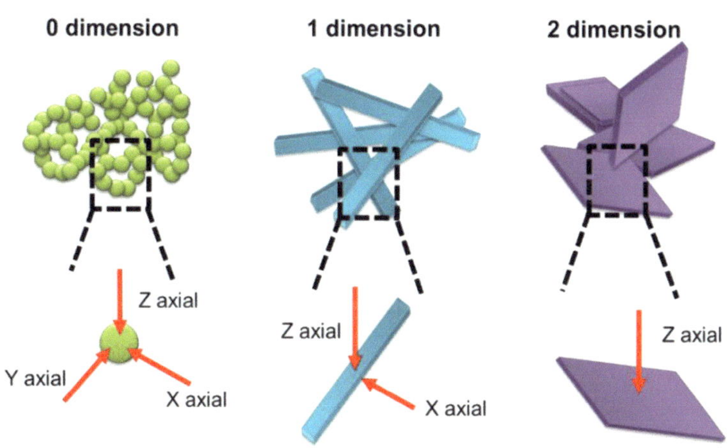

Figure 3.
The scheme of zero-dimensional (0D), one-dimensional (1D), and two-dimensional (2D) structure.

expected to supply the high electrochemical surface area, and 1D/2D structures are claimed to have directional electron transfer pathways. In this chapter, different strategies of designing nanostructured carbon materials, conductive polymers, and transition metal compounds to increase their active surface area/charge transfer route will be systematically discussed. The corresponding DSSC performance is also included.

2. Nanostructure materials of counter electrode

2.1 Carbon materials

Carbon materials, composed of carbon atoms having sp^2-hybridization (e.g., graphite, graphene, graphene oxide, and carbon nanotube) or sp^3-hybridization (e.g., carbon black and activated carbon), have been widely investigated as the CEs due to their material specialties such as low cost, high electrical conductivity, high thermal stability, and good corrosion resistance [3, 4, 80–84]. The carbon materials with sp^2-hybridization normally exhibit a 1D or 2D structure, and those with sp^3-hybridization have a 0D structure. However, carbon materials often exhibit serious aggregation that reduces the material conductivity and limits the electrochemical surface area. Here we include several studies which provided practical strategies to overcome this problem and reach comparable/better cell efficiencies than the Pt-incorporated DSSCs, as listed in **Table 1**.

For example, Fan et al. used a small 0D porous carbon nanoball (diameter = 20 ± 3 nm) to assemble a large 0D hollow nanoball (diameter = 100 ± 10 nm), as shown in **Figure 4(a)** [12]. Tseng et al. introduced a one-step synthetic method to make tens of 2D nitrogen-doped graphene with a thickness of ~3.5 nm stacking together to form a building block as a 0D hollow nanoball, as shown in **Figure 4(b)** [22]. Fan et al. used a small 0D porous carbon nanoball (diameter = 20 ± 3 nm) to assemble a large 0D hollow nanoball (diameter = 100 ± 10 nm), as shown in **Figure 4(b)** [12]. The hollow nanoballs consisting of nitrogen-doped graphene and porous carbon gave their DSSCs ηs of 7.53 and 8.67%, respectively, which were comparable to the Pt-based cells. Besides, nitrogen-doped graphene (**Figure 4(c)**, 7.07%) [13] and the wrinkled 2D graphene (**Figure 4(d)**, 7.80%) [14] nanosheets were used to form a honeycomb-like structure having extra surface area and vertically aligned sub-nanosheets as the additional electron transfer routes.

Materials	η (%)	η of Pt (%)	Structure	Ref
Porous carbon	8.67	9.34	Hollow nanoball	[12]
Nitrogen-doped graphene	7.53	7.70	Hollow nanoball	[22]
Nitrogen-doped graphene	7.07	7.44	Honeycomb	[13]
Graphene	7.80	8.00	Honeycomb	[14]
Carbon nanotubes and graphene	8.2	6.4	Nanotube vertically fused onto the nanosheet	[15]
Carbon nanotube and N-doped graphene	8.31	7.56	Nanotube intertwined with nanosheet	[16]
Carbon nanotube and graphene oxide	6.91	7.26	Nanotube embedded in nanosheet	[27]

Table 1.
A partial list of literature on the DSSCs with carbon material-based CEs.

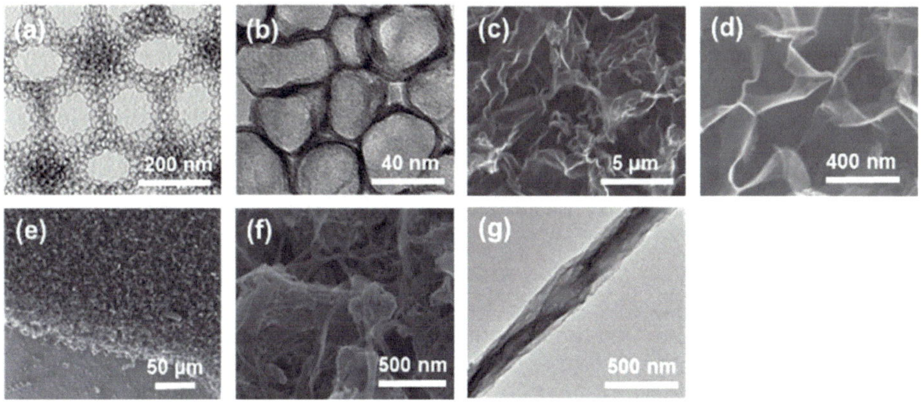

Figure 4.
The structures of carbon materials including (a and b) hollow nanoball [12, 22], (c) nanosheet [13], (d) honeycomb [14], and (e–g) nanotube with nanosheet [15, 16, 27].

The combination of few kinds of carbon materials was reported to form a hierarchical structure, which could not only create a high surface area but also a directional electron transfer pathway. Dong et al. made 1D few-walled carbon nanotubes (CNTs, tens of microns long) vertically fuse onto the 2D graphene nanosheet (<1 nm thick), as shown in **Figure 4(e)** [15]. The red seven-membered rings at the neck seamlessly fuse the tubular CNTs to the planar graphene without obvious CNT aggregation (**Figure 5(a)**). Even though the CNTs only covered few parts of the electrode surface, they still benefited the electrolyte wetting and electron transfer rate within the counter electrode, leading to a better η (8.2%) than the Pt-based cell (6.4%). Ma et al. reported a similar hierarchical structure, where the single-walled carbon nanotube (SWCNT) was located on a flat N-doped graphene (N-doped GN) nanosheet by the z axis direction (**Figure 4(f)**) [16]. As a result, composite SWCNT@N-doped GN reached higher η (8.31%) than Pt (7.56%). Yeh et al. prepared a hybrid heterostructure of multiwalled carbon nanotube (MWCNT, diameter = 15 nm) and reduced graphene oxide (rGO, thickness = 25 nm) nanosheet, where the 2D rGO nanosheet-like shell was wrapped around the 1D CNT core (**Figure 4(g)**) [27]. In the MWCNT@rGO composite material, the tubular MWCNTs functioned as the 1D heterogeneous electron transfer pathways, which provided sufficient electrons to the electrochemical reaction; at the same time, the 2D rGO nanosheet supplied multiple edges as the active sites to reduce I_3^- to I^- (**Figure 5(b)**). The hybrid heterostructure of MWCNT@rGO was found to avoid

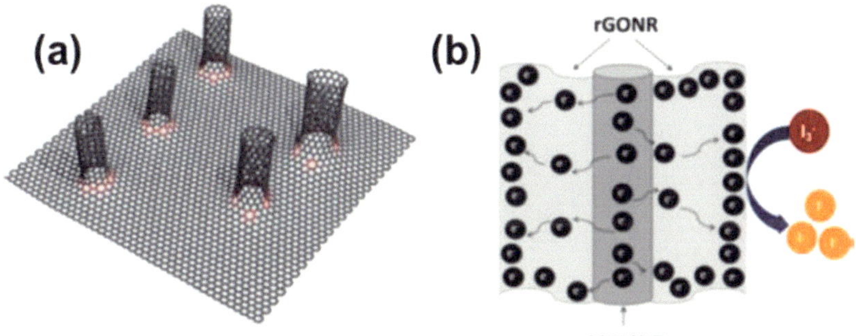

Figure 5.
The scheme of hierarchical structures of (a and b) nanotube with nanosheet [15, 27].

the aggregations among the MWCNTs or among the rGOs. Thus, the MWCNT@ rGO rendered its DSSC an η of 6.91%, which is close to the Pt-based cell (7.26%).

2.2 Conductive polymer materials

Since 2000, the conductive polymer material has been discovered by Shirakawa, MacDiarmid, and Heeger [29]. Conductive polymers have attracted much attention as DSSC CEs owing to their excellent conductivity, good adhesion to the substrate, easy fabrication, light-weight, and good accessibility in terms of roll-to-roll processing. Common conductive polymers include poly(3,4-ethylenedioxythiophene) (PEDOT) [30, 35, 36, 40–43], poly(3,4-ethylenedioxythiophene)-poly(styrene sulfonate) (PEDOT:PSS) [30, 43, 48], poly(hydroxymethyl 3,4-ethylene-dioxythiophene) (PEDOT-MeOH) [46], poly(3,4-propylenedioxythiophene) (PProDOT) [85, 86], poly(3,3-diethyl-3,4-dihydro-2H-thieno-[3,4-b][1,4]dioxepine) (PProDOT-Et_2) [85–87], poly(2,2-dimethyl-3,4-propylenedioxythiophene) (PProDOT-Me_2) [85], polythiophene (PT) [33], sulfonated poly(thiophene-3-[2-(2-methoxyethoxy) ethoxy]-2,5-diyl) (s-PT) [48], polyaniline (PANI) [34], and polypyrrole (PPy) [31, 38, 39, 44, 48], and their molecular structures are shown in **Figure 6**. However, conductive polymers often showed a flat or a mesoporous structure, meaning their lacks of the directional electron transfer pathways. Because of the synthetic difficulties, very few conductive polymers can form a 0D/1D/2D structure, as listed in **Table 2**.

The hierarchical nanosphere with PPy (denoted PPy-HNS) has the hierarchical nanospherical structure with an average diameter of 100–200 nm, as shown in **Figure 7(a)** [44]. The PPy-HNS has the following photovoltaic parameters: a V_{OC} of 0.70 V, a J_{SC} of 16.49 mA cm^{-2}, a FF of 0.58, and an η of 6.71%. The nanopatterning process is one of the methods to obtain the specific structure [42]. The nanopatterned PEDOT CE shows uniform hole patterns with ~100 nm diameter (**Figure 7(b)**) and exhibits an η of 6.71%. Regardless of the nanoparticle or nanosize hole, their corresponding η values are still lower than the η of Pt CE. Therefore, there are other structures, which were synthesized to overcome the challenge.

The flexible PPy membrane is composed of nanotubes that are about 50 nm in diameter, as shown in **Figure 7(c)** [38]. The paper-like PPy membranes exhibit

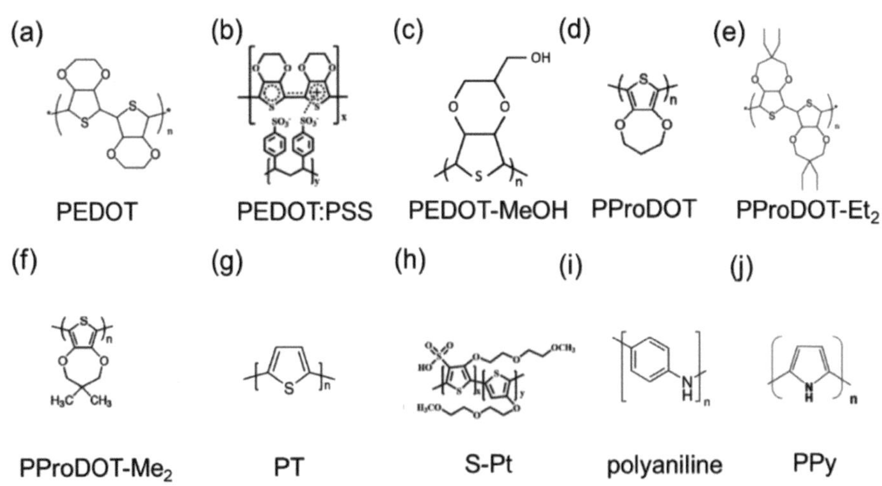

Figure 6.
The molecular structures of (a) PEDOT, (b) PEDOT:PSS [30, 43, 48], (c) PEDOT-MeOH [46], (d) PProDOT [85, 86], (e) PProDOT-Et_2 [85–87], (f) PProDOT-Me_2 [85], (g) PT [33], (h) s-PT [48], (i) PANI [34], and (j) polypyrrole (PPy) [31, 38, 39, 44, 48].

Materials	η (%)	η of Pt (%)	Structure	Ref
PPy	6.71	7.47	Hierarchical nanosphere	[44]
PEDOT	7.10	7.60	Nanosized hole	[42]
PPy	5.27	6.25	Nanotubes	[38]
PEDOT-MeOH	9.13	8.94	Tube-coral array	[46]
PEDOT	8.10	8.26	Nanorod	[35]
PANI	6.86	7.21	Nanorod	[34]
PPy	6.80	7.80	Nanosheet	[39]

Table 2.
A partial list of literature on the DSSCs with conductive polymer material-based CEs.

ηs of 5.27%, which is about 84% of the cell performance with a conventional Pt/FTO CE (6.25%). Usually, the rod structure of PEDOT is obtained by a template method. The PEDOT-MeOH tube-coral array is synthesized by a template-free and bottom-up electropolymerization technique [46]. The PEDOT-MeOH TCA shows three advantages: (1) an enhanced conjugation of the PEDOT main chain due to the electron-donating MeOH group, (2) fast one-dimensional charge transfer routes, and (3) extended electroactive sites. The PEDOT-MeOH TCA has a highly porous surface with an average length of 5 mm and an average diameter of 500 nm, as shown in **Figure 7(d)**. Besides, PEDOT-MeOH TCA has vertically grown on their FTO substrates. The PEDOT-MeOH TCA CE, the co-existence of the 1D charge transfer pathways, and large active surface area on the PEDOT-MeOH TCA give its DSSC an η of 9.13%, which is higher than the Pt CE (8.94%). This performance is rarely found in pure conductive polymer materials. The patterned PEDOT with rod is obtained by the nanopatterning procedure. It has the height of 67 nm and width of 100 nm, as shown in **Figure 7(e)** [35]. The DSSCs of patterned PEDOT exhibits an η of 8.10%, which is close to Pt CE (8.26%). The prickly polyaniline nanorods (PPNR) display a prickly nanorod structure with the diameter of ~80 nm and the length of several micrometers, as shown in **Figure 7(f)**. The PPNR CE exhibits an η of 6.86%. The ultrathin polypyrrole nanosheets (UPNSs) have a nanoscale thickness of 50 nm and sheet morphologies by using a sodium decylsulfonate template, as shown in **Figure 7(g)**. The DSSC using UPNS CE shows an η of 6.80%.

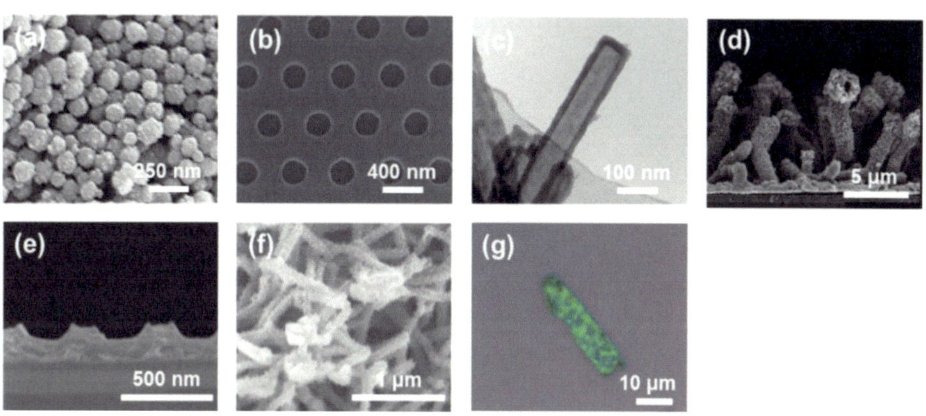

Figure 7.
The structures of conductive polymer materials include (a) hierarchical nanosphere of PPy [44], (b) nanosized hole of PEDOT [42], (c) nanotube of PPy [38], (d) hollow tubular structure of PEDOT-MeOH [46], (e) nanorod of PEDOT [35], (f) nanorod of PANI [34], and (g) nanosheet of PPy [39].

2.3 Transition metal composites

Transition metal composites (TMC) possess high potential to replace Pt CE in DSSCs because of the similar electronic structures between TMCs and Pt. Metal compounds, including carbides, nitrides, chalcogenides, oxides, phosphides, and so on, have been applied as an electrocatalyst in DSSCs to replace expensive Pt. It is still a challenge to replace Pt with TMCs due to the relatively low conductivity of TMCs. Accordingly, various TMC structures, including nanoparticle, hollow sphere, nanorod array, nanowall, hierarchical nanorod, etc., are investigated to improve the performance of TMC-based CEs, as shown in **Figure 8**. The corresponding ηs of DSSCs with various structures are listed in **Table 3**.

α-NiS has a sphere-like morphology with a diameter of 50–80 nm, as shown in **Figure 8(a)** [78]. The other NiS (β-NiS) has a nanorod 2–5 μm in length and 1000 nm in diameter. The DSSC of α-NiS CE has a better η (5.20%) than the β-NiS (η of 4.20%). The particle size of α-NiS is much smaller than nanorods of β-NiS. The smaller the size of a particle, the larger specific surface area it possesses. With the increase of specific surface area of α-NiS, the conversion efficiency reaches a higher value. The nanoparticle of $CoSe_2/CoSeO_3$ ($CoSe_2/CoSeO_3$-NP) has a diameter of 50–60 nm, as shown in **Figure 8(b)** [74]. And $CoSe_2/CoSeO_3$-NP has a larger reaction area than the nanorod and nanocube of $CoSe_2/CoSeO_3$, confirmed by the electrochemical double-layer capacitance, which is positively related to the

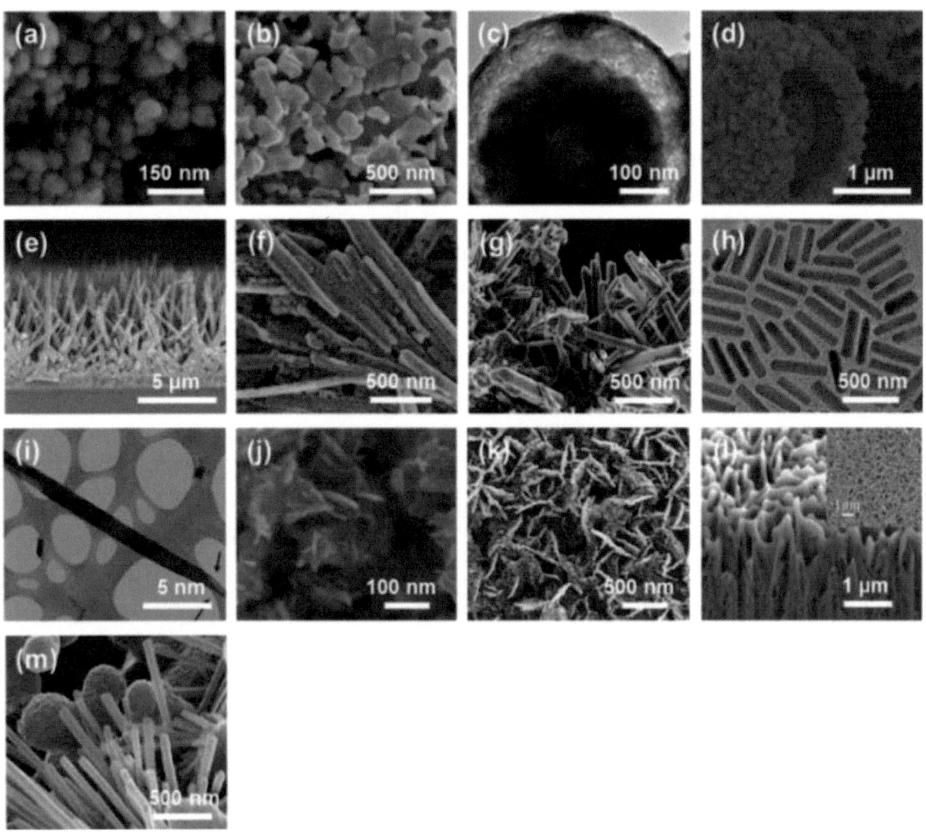

Figure 8.
The structures of transition metal materials including (a and b) nanoparticle [78], (c) double-shelled ball-in-ball hollow sphere [70, 74], (d) hollow spherical particle [71], (e) acicular nanorod array [49], (f–h) nanorod [53, 54, 75], (i and j) nanosheet [55, 66], (k and l) nanowall [64, 72], and (m) hierarchical nanosphere with nanorod [77].

Materials	η (%)	η of Pt (%)	Structure	Ref
NiS	5.20	6.30	Nanoparticle	[78]
$CoSe_2/CoSeO_3$	9.27	7.91	Nanoparticle	[74]
$NiCo_2S_4$	9.49	8.30	Double-shelled ball-in-ball hollow sphere	[70]
$NiCo_{0.2}$@C	9.30	8.04	Hollow spherical particle	[71]
CoS	7.67	7.70	Acicular nanorod array	[49]
MoN	7.29	7.42	Nanorod	[54]
$CoSe_2$	10.20	8.17	Nanorod	[53]
Ni_3S_4-Pt_2Fe_1	8.79	7.83	Nanorod	[75]
$NbSe_2$	7.73	7.01	Nanosheet	[55]
WSe_2	7.48	7.91	Nanosheet	[66]
$CoSe_2$	8.92	8.25	Nanoclimbing wall	[64]
$Cu_xZn_ySn_zS$	7.44	7.21	Nanowall	[72]
$TiO_{1.1}Se_{0.9}$	9.47	7.75	Nanosphere and nanorod	[77]

Table 3.
A partial list of literature studies on the DSSCs with conductive polymer material-based CEs. The dye of DSSCs is N719.

reaction area. The DSSC of $CoSe_2/CoSeO_3$-NP CE has an η of 9.27%, which is better than those of the nanorod and nanocube of $CoSe_2/CoSeO_3$ and is higher than that of the Pt CE (7.91%). The double-shelled hollow sphere (BHSs) structure exists in $NiCo_2S_4$ BHSs with the separation of hollow and solid parts. In **Figure 8(c)** [70], the diameter of the inner shell is 300 nm and that of the outer shell is approximately 550 nm. The thickness of the outer thin shell is 10–30 nm, which is quite less than that of the inner shell. The DSSC of $NiCo_2S_4$ BHSs CE exhibits an η of 9.49%, which is higher than that of the Pt CE (η of 8.30%). From a broken $NiCo_{0.2}$@C microsphere shown in **Figure 8(d)** [71], the well-defined hollow structure with a shell thickness of around 200 nm can be observed. Meaningfully, the hollow spherical space can greatly shorten the diffusion paths within the electrode and serves as a robust reservoir for ions. The $NiCo_{0.2}$@C exhibits an η of 9.30%, which is higher than that of the Pt CE (η of 8.04%). Most of the nanoparticles, double-shelled ball-in-ball hollow sphere, and hollow spherical particle structures have better ηs than the Pt CE.

Although TMCs present good electrocatalytic ability, the electrons may be insufficient at active sites. The rod structure is claimed to provide the specific electron transfer. It can supply sufficient electrons to keep consistent electrocatalytic reaction. From **Figure 8(e)**, it can be observed that CoS has 1D acicular nanorod arrays with the relatively rough surface of the nanorods (noted CoS ANRAs-24h) [49]. It is vertical to the FTO substrate and has a height of about 7 μm. The DSSC with CoS ANRAs-24h CE shows an η of 7.67%, which is virtually the same as the sputtered Pt-CE (η of 7.70%). The MoN nanorod (NR) on the Ti substrate has a one-dimensional structure with a diameter of 40–100 nm and a length of 0.5–2 mm, as shown in **Figure 8(f)** [54]. The electrode structure is expected to trigger positive effects on the electrochemical processes occurring in the electrode films. The MoN NR-Ti CE shows comparable performance to that using a Pt-FTO glass electrode with a V_{OC} of 0.740 V, a J_{SC} of 15.26 mA cm^{-2}, a FF of 0.65, and an η of 7.29%. The single-crystal $CoSe_2$ has nanorods 50–800 nm in length and 20–150 nm in width, as shown in **Figure 8(g)**, and possesses a lattice spacing of 3.71 ± 0.01 Å, corresponding to the

(110) planes of orthorhombic $CoSe_2$ [53]. Impressively, the single-crystal $CoSe_2$ CE produces an η of 10.20% with a V_{OC} of 0.753 V, a J_{SC} of 18.55 mA/cm^{-2}, and a FF of 0.73, which is better than the Pt CE (8.17%). The Ni_3S_4-PtFe heteronanorods are highly monodispersed with an average length of ~34.0 nm and an average diameter of 9.0 nm, as shown in **Figure 8(h)** [75]. The DSSCs using Ni_3S_4-PtFe produce an η of 8.79%, which is higher than that of the Pt CE (7.83%). The 1-D structure is obviously promoting the electrocatalytic ability of TMCs and most of the 1-D structures for TMCs exhibit a better η than Pt CE. It can be claimed that the 1-D structures of TMCs could replace the Pt CE.

The 2D structure of TMCs also has a specific electron pathway and it could be vertical to the substrate to offer sufficient electrons on active sites. Moreover, the hierarchical structure has both the advantages of a large reaction area and vertical electron pathway. For example, the direction of the fractured $NbSe_2$ sheet shows a structure with the [001] crystallographic orientation and revealed a very thick (>100 mm), disordered network arrangement of 2D sheets, as shown in **Figure 8(i)**; in comparison, the ground materials were very thin, separated nanosheets [55]. The NbSe2 sheet CE has an η of 7.73%, which reveals the potential to replace Pt CE (7.01%). The WSe_2 is composed of several interlaced nanosheets with an average thickness of approximately 15 nm and a width between 60 and 100 nm, as shown in **Figure 8(j)** [66]. The WSe_2 CE shows good electrical conductivity, subsequent energy band calculation results, and large reaction area that exhibits an η of 7.48%.

Vertically-aligned structures of electrocatalysts were reported to facilitate faster charge transport from the substrate through the electrocatalysts to the electrolyte [64, 72, 77], as shown in **Figure 9**. This structure is expected to have better electrocatalytic ability. The nanowall and the hierarchical nanorod are used with TMCs. The $CoSe_2$ nanoclimbing wall ($CoSe_2$/C-NCW) reveals arrays of vertically-aligned nanowalls with sharp edges, as shown in **Figure 8(k)** [64]. In addition, the nanowalls are covered with dot-matrix-like projections; these projections are expected to provide a large surface area to the film. On account of direct electron transfer and large surface area, the $CoSe_2$/C-NCW film, on the whole, could be a better electrocatalyst for the reduction of I_3^- to I^-, as shown in **Figure 9(a)**. The cell with $CoSe_2$/C-NCW CE reaches the highest efficiency of 8.92%, with a V_{OC} of 0.73 V, a J_{SC} of 18.03 mA cm^{-2}, and an FF of 0.67; this efficiency is even higher than that of the cell with Pt (8.25%). The CZTS nanowall electrodes (NWD) on Mo substrate show nanowalls with a width of ~500 nm, a thickness of nearly 15 nm, and a height of ~1.5 μm, which were adequately aligned in a densely packed array, which was nearly perpendicular to the surface of the Mo substrate, as shown in **Figure 8(l)** [72]. In this case, CZTS-NWD demonstrates a concept of "nano-geogrid"-reinforced CZTS nanowall electrode by synthesizing a thin layer of a porous CZTS nanostructure mimicking a geogrid on a substrate and then fabricating a CZTS nanowall on top of the nanostructure, as shown in **Figure 9(b)**. The η of the NWD device is 7.44%, which is comparable to the Pt device (7.21%). **Figure 8(m)** shows the film of

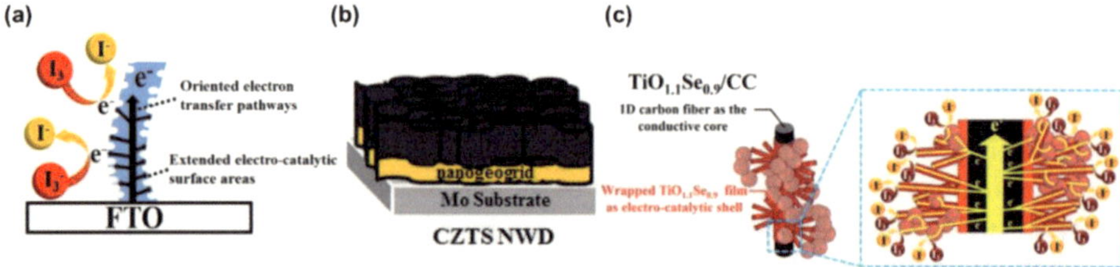

Figure 9.
The scheme of vertical aligned structures of (a and b) nanowall and (c) nanosphere with nanorod [64, 77].

$TiO_{1.1}Se_{0.9}$ wrapping around a carbon fiber, and many nanospheres and nanorods of $TiO_{1.1}Se_{0.9}$ are grown on the $TiO_{1.1}Se_{0.9}$ under-layer [77]. The $TiO_{1.1}Se_{0.9}$ nanorods are perpendicular to the surface of the carbon fiber; $TiO_{1.1}Se_{0.9}$ nanorods, therefore, are expected to facilitate a fast and 1D electron transport from the CC substrate to the interface with the electrolyte, where I_3^- reduction occurs The correctional images show the hierarchical electron transfer route electrode, where the carbon fiber in the CC serves for transporting the main stream of electrons through its 1D direction, and the deposited electrocatalysts provide secondary channels of electrons for executing the reduction of I_3^-, as shown in **Figure 9(c)**. The DSSC with the $TiO_{1.1}Se_{0.9}$–3/CC shows the best performance and gives the best η of 9.47%. Furthermore, the DSSC is used for further comparison with those cells with the Pt/CC (7.75%). In conclusion, the TMCs have good electrocatalytic ability and possess vertically aligned structures. They exhibit great ηs, which are better than that of the Pt CE.

3. Summary and future prospects

The counter electrode is a paramount part of DSSCS and has a significant influence on both the photovoltaic performance and the device cost of DSSCs. As a counter electrode, it must possess high conductivity and good catalytic activity toward electrolyte regeneration, as well as good stability. The DSSC devices employing CEs of different materials including carbon materials, conductive polymers, and transition metal composites have been summarized and discussed. One key point is that the CE performance can be optimized by combining special nanostructures into CE films to promote the industrialization of Pt-free CE catalysts. The nanostructure can briefly be classified into 0D, 1D, and 2D, which have different properties. The different materials with various nanostructures can overcome the problem of the material.

The carbon materials have numerous advantages including low cost, plasticity, simple fabrication procedures, high electrical conductivity, high thermal stability, and good corrosion resistance. The η of carbon materials has been improved by the hierarchal structures (nanotube with nanosheet and nanotube with nanoribbon) and most of the carbon materials with hierarchal structure CE have a better value of η than the traditional Pt CE. However, most of the performances of the DSSCs with carbon material CEs are still slightly lower than those DSSCs with Pt CEs. This mostly results from various resistances associated with the structurally complex carbon electrodes, such as bulk resistance through the comparatively thick carbon CE, contact resistance to the TCO substrate, the diffusion resistance in the pores of the CE, etc.

The conductive polymer materials possess outstanding electron conductivity, good adhesion, and easy fabrication. According to the literature above, it can be concluded that the 1D structure conductive polymer material-based CE can provide better η than the particles, nanosphere, and nanosheet structures. Although conductive polymer materials have larger reaction area and specific electron pathway, most of the conductive polymer material-based CEs still have a lower η than the Pt-based CEs. Only a few examples show better performance than Pt CE. It means that the conductive polymer materials need a hybrid with other electrocatalysts to obtain better electrocatalytic ability.

The TMCs exhibit great electrocatalytic ability, easy preparation, and modification. However, the poor conductivity needs to be solved in order to replace Pt CE. By synthesizing nanostructures, including nanoparticle, double-shelled ball-in-ball hollow sphere, hollow spherical particle, acicular nanorod array, nanorod, nanosheet, nanoclimbing wall, hierarchical nanorod, etc., TMCs reveal better performances than the Pt CE. It can be said that the TMCs with nanostructure successfully replace Pt CE.

Moreover, changing DSSC electrolyte toward the Cu, Co, Fe, etc. redox couples is another important research topic. Furthermore, the dim light condition application is another prospect of DSSCs. Among them, the matching material of CE is a key point to promise the η of DSSC. To get the point, it is dependent on the nanostructure and the hybrid materials, such as carbon materials with TMCs, carbon materials with conductive polymer materials, conductive polymer materials with TMCs, and TMCs with both conductive polymer materials and carbon material. Further development should focus on these main requirements: conductivity, catalytic activity, stability, efficiency, cost, and environmental friendliness. Also, the regulation mechanism for photo-induced charge carrier generation, evolution, and transportation should be of concern.

Acknowledgements

This work was supported by the Ministry of Science and Technology (MOST) of Taiwan. Professor Chuan-Pei Lee especially thanks the financial support of MOST, under grant numbers 107-2113-M-845-001-MY3.

Author details

Yi-June Huang[1], Han-Ting Chen[1], Shiuan-Bai Ann[1], Chun-Ting Li[2] and Chuan-Pei Lee[3*]

1 Department of Chemical Engineering, National Taiwan University, Taiwan

2 Institute of Chemistry, Academia Sinica, Taiwan

3 Department of Applied Physics and Chemistry, University of Taipei, Taiwan

*Address all correspondence to: cplee@utaipei.edu.tw

References

[1] Wu M, Ma T. Platinum-free catalysts as counter electrodes in dye-sensitized solar cells. ChemSusChem. 2012;**5**: 1343-1357. DOI: 10.1002/cssc.201100676

[2] Bella F, Gerbaldi C, Barolo C, Gratzel M. Aqueous dye-sensitized solar cells. Chemical Society Reviews. 2015;**44**:3431-3473. DOI: 10.1039/c4cs00456f

[3] Kouhnavard M, Ludin NA, Ghaffari BV, Sopian K, Ikeda S. Carbonaceous materials and their advances as a counter electrode in dye-sensitized solar cells: Challenges and prospects. ChemSusChem. 2015;**8**: 1510-1533. DOI: 10.1002/cssc.201500004

[4] Roy-Mayhew JD, Aksay IA. Graphene materials and their use in dye-sensitized solar cells. Chemical Reviews. 2014;**114**:6323-6348. DOI: 10.1021/cr400412a

[5] Lee CP, Lin CA, Wei TC, Tsai ML, Meng Y, Li CT, et al. Economical low-light photovoltaics by using the Pt-free dye-sensitized solar cell with graphene dot/PEDOT:PSS counter electrodes. Nano Energy. 2015;**18**:109-117. DOI: 10.1016/j.nanoen.2015.10.008

[6] Ganesan P, Yella A, Holcombe TW, Gao P, Rajalingam R, Al-Muhtaseb SA, et al. Unravel the impact of anchoring groups on the photovoltaic performances of diketopyrrolopyrrole sensitizers for dye-sensitized solar cells. ACS Sustainable Chemistry & Engineering. 2015;**3**:2389-2396. DOI: 10.1021/acssuschemeng.5b00332

[7] Chen CY, Jian ZH, Huang SH, Lee KM, Kao MH, Shen CH, et al. Performance characterization of dye-sensitized photovoltaics under indoor lighting. Journal of Physical Chemistry Letters. 2017;**8**:1824-1830. DOI: 10.1021/acs.jpclett.7b00515

[8] Chou HH, Liu YC, Fang G, Cao QK, Wei TC, Yeh CY. Structurally simple and easily accessible perylenes for dye-sensitized solar cells applicable to both 1 Sun and dim-light environments. ACS Applied Materials & Interfaces. 2017;**9**:37786-37796. DOI: 10.1021/acsami.7b11784

[9] Tsai MC, Wang CL, Chang CW, Hsu CW, Hsiao YH, Liu CL, et al. A large, ultra-black, efficient and cost-effective dye-sensitized solar module approaching 12% overall efficiency under 1000 lux indoor light. Journal of Materials Chemistry A. 2018;**6**: 1995-2003. DOI: 10.1039/c7ta09322e

[10] Wu M, Ma T. Recent Progress of counter electrode catalysts in dye-sensitized solar cells. The Journal of Physical Chemistry C. 2014;**118**: 16727-16742. DOI: 10.1021/jp412713h

[11] Hou Y, Wang D, Yang XH, Fang WQ, Zhang B, Wang HF, et al. Rational screening low-cost counter electrodes for dye-sensitized solar cells. Nature Communications. 2013;**4**:1583. DOI: 10.1038/ncomms2547

[12] Fan SQ, Fang B, Kim JH, Jeong B, Kim C, Yu JS, et al. Ordered multimodal porous carbon as highly efficient counter electrodes in dye-sensitized and quantum-dot solar cells. Langmuir. 2010;**26**:13644-13649. DOI: 10.1021/la1019873

[13] Xue Y, Liu J, Chen H, Wang R, Li D, Qu J, et al. Nitrogen-doped graphene foams as metal-free counter electrodes in high-performance dye-sensitized solar cells. Angewandte Chemie (International Ed. in English). 2012;**51**:12124-12127. DOI: 10.1002/anie.201207277

[14] Wang H, Sun K, Tao F, Stacchiola DJ, Hu YH. 3D honeycomb-like structured graphene and its high efficiency as a

counter-electrode catalyst for dye-sensitized solar cells. Angewandte Chemie (International Ed. in English). 2013;**52**:9210-9214. DOI: 10.1002/anie.201303497

[15] Dong P, Zhu Y, Zhang J, Hao F, Wu J, Lei S, et al. Vertically aligned carbon nanotubes/graphene hybrid electrode as a TCO- and Pt-free flexible cathode for application in solar cells. Journal of Materials Chemistry A. 2014;**2**: 20902-20907. DOI: 10.1039/c4ta05264a

[16] Ma J, Li C, Yu F, Chen J. 3 D single-walled carbon nanotube/graphene aerogels as pt-free transparent counter electrodes for high efficiency dye-sensitized solar cells. ChemSusChem. 2014;**7**:3304-3311. DOI: 10.1002/cssc.201403062

[17] Zheng X, Deng J, Wang N, Deng D, Zhang WH, Bao X, et al. Podlike N-doped carbon nanotubes encapsulating FeNi alloy nanoparticles: High-performance counter electrode materials for dye-sensitized solar cells. Angewandte Chemie (International Ed. in English). 2014;**53**:7023-7027. DOI: 10.1002/anie.201400388

[18] Wang CC, Lu SY. Carbon black-derived graphene quantum dots composited with carbon aerogel as a highly efficient and stable reduction catalyst for the iodide/tri-iodide couple. Nanoscale. 2015;**7**:1209-1215. DOI: 10.1039/c4nr06118g

[19] Yang W, Ma X, Xu X, Li Y, Raj SI, Ning G, et al. Sulfur-doped porous carbon as metal-free counter electrode for high-efficiency dye-sensitized solar cells. Journal of Power Sources. 2015;**282**:228-234. DOI: 10.1016/j.jpowsour.2015.02.060

[20] Zhang J, Yu M, Li S, Meng Y, Wu X, Liu J. Transparent conducting oxide-free nitrogen-doped graphene/reduced hydroxylated carbon nanotube composite paper as flexible counter electrodes for dye-sensitized solar cells. Journal of Power Sources. 2016;**334**:44-51. DOI: 10.1016/j.jpowsour.2016.10.012

[21] Memon AA, Arbab AA, Sahito IA, Sun KC, Mengal N, Jeong SH. Synthesis of highly photo-catalytic and electro-catalytic active textile structured carbon electrode and its application in DSSCs. Solar Energy. 2017;**150**:521-531. DOI: 10.1016/j.solener.2017.04.052

[22] Tseng CA, Lee CP, Huang YJ, Pang HW, Ho KC, Chen YT. One-step synthesis of graphene hollow nanoballs with various nitrogen-doped states for electrocatalysis in dye-sensitized solar cells. Materials Today Energy. 2018;**8**: 15-21. DOI: 10.1016/j.mtener.2018.02.006

[23] Lee CP, Lai KY, Lin CA, Li CT, Ho KC, Wu CI, et al. A paper-based electrode using a graphene dot/PEDOT:PSS composite for flexible solar cells. Nano Energy. 2017;**36**:260-267. DOI: 10.1016/j.nanoen.2017.04.044

[24] Li CT, Lee CT, Li SR, Lee CP, Chiu IT, Vittal R, et al. Composite films of carbon black nanoparticles and sulfonated-polythiophene as flexible counter electrodes for dye-sensitized solar cells. Journal of Power Sources. 2016;**302**:155-163. DOI: 10.1016/j.jpowsour.2015.10.028

[25] Li YY, Li CT, Yeh MH, Huang KC, Chen PW, Vittal R, et al. Graphite with different structures as catalysts for counter electrodes in dye-sensitized solar cells. Electrochimica Acta. 2015;**179**:211-219. DOI: 10.1016/j.electacta.2015.06.007

[26] Chen PY, Li CT, Lee CP, Vittal R, Ho KC. PEDOT-decorated nitrogen-doped graphene as the transparent composite film for the counter electrode of a dye-sensitized solar cell. Nano Energy. 2015;**12**:374-385. DOI: 10.1016/j.nanoen.2015.01.010

[27] Yeh MH, Lin LY, Sun CL, Leu YA, Tsai JT, Yeh CY, et al. Multiwalled

carbon nanotube@reduced graphene oxide nanoribbon as the counter electrode for dye-sensitized solar cells. The Journal of Physical Chemistry C. 2014;**118**:16626-16634. DOI: 10.1021/jp412542d

[28] Lin CA, Lee CP, Ho ST, Wei TC, Chi YW, Huang KP, et al. Nitrogen-doped Graphene/platinum counter electrodes for dye-sensitized solar cells. ACS Photonics. 2014;**1**:1264-1269. DOI: 10.1021/ph500219r

[29] Groenendaal_et_al-2000-Advanced_Materials.pdf. DOI: 10.1002/(SICI)1521-4095(200004)12:73.3.CO;2-3

[30] Xia J, Masaki N, Jiang K, Yanagida S. The influence of doping ions on poly(3,4-ethylenedioxythiophene) as a counter electrode of a dye-sensitized solar cell. Journal of Materials Chemistry. 2007;**17**:2845. DOI: 10.1039/b703062b

[31] Xia J, Chen L, Yanagida S. Application of polypyrrole as a counter electrode for a dye-sensitized solar cell. Journal of Materials Chemistry. 2011;**21**:4644. DOI: 10.1039/c0jm04116e

[32] Zhang R, Li Z, Xu J, Xia J. Synthesis and application of poly(bis-3,4-ethylenedioxythiophene methine)s as novel counter electrodes in dye-sensitized solar cells. Solar Energy. 2018;**173**:1189-1196. DOI: 10.1016/j.solener.2018.08.063

[33] Wu J, Lan Z, Lin J, Huang M, Huang Y, Fan L, et al. Counter electrodes in dye-sensitized solar cells. Chemical Society Reviews. 2017;**46**:5975-6023. DOI: 10.1039/c6cs00752j

[34] Wang G, Yan C, Zhang W. Prickly polyaniline nano/microstructures as the efficient counter electrode materials for dye-sensitized solar cells. Journal of Nanoparticle Research. 2017;**19**:395. DOI: 10.1007/s11051-017-4087-0

[35] Park C, Na J, Kim E. Cross stacking of nanopatterned PEDOT films for use as soft electrodes. ACS Applied Materials & Interfaces. 2017;**9**:28802-28809. DOI: 10.1021/acsami.7b07799

[36] Li H, Xiao Y, Han G, Hou W. Honeycomb-like poly(3,4-ethylenedioxythiophene) as an effective and transparent counter electrode in bifacial dye-sensitized solar cells. Journal of Power Sources. 2017;**342**:709-716. DOI: 10.1016/j.jpowsour.2017.01.007

[37] Yun S, Hagfeldt A, Ma T. Pt-free counter electrode for dye-sensitized solar cells with high efficiency. Advanced Materials. 2014;**26**:6210-6237. DOI: 10.1002/adma.201402056

[38] Peng T, Sun W, Huang C, Yu W, Sebo B, Dai Z, et al. Self-assembled free-standing polypyrrole nanotube membrane as an efficient FTO- and Pt-free counter electrode for dye-sensitized solar cells. ACS Applied Materials & Interfaces. 2014;**6**:14-17. DOI: 10.1021/am404265q

[39] Hwang DK, Song D, Jeon SS, Han TH, Kang YS, Im SS. Ultrathin polypyrrole nanosheets doped with HCl as counter electrodes in dye-sensitized solar cells. Journal of Materials Chemistry A. 2014;**2**:859-865. DOI: 10.1039/c3ta13367b

[40] Carli S, Casarin L, Bergamini G, Caramori S, Bignozzi CA. Conductive PEDOT covalently bound to transparent FTO electrodes. The Journal of Physical Chemistry C. 2014;**118**:16782-16790. DOI: 10.1021/jp412758g

[41] Kwon J, Park NG, Lee JY, Ko MJ, Park JH. Highly efficient monolithic dye-sensitized solar cells. ACS Applied Materials & Interfaces. 2013;**5**:2070-2074. DOI: 10.1021/am302974z

[42] Kwon J, Ganapathy V, Kim YH, Song KD, Park HG, Jun Y, et al.

Nanopatterned conductive polymer films as a Pt, TCO-free counter electrode for low-cost dye-sensitized solar cells. Nanoscale. 2013;5:7838-7843. DOI: 10.1039/c3nr01294h

[43] Zhang TL, Chen HY, Su CY, Kuang DB. A novel TCO- and Pt-free counter electrode for high efficiency dye-sensitized solar cells. Journal of Materials Chemistry A. 2013;**1**: 1724-1730. DOI: 10.1039/c2ta00974a

[44] Chang LY, Li CT, Li YY, Lee CP, Yeh MH, Ho KC, et al. Morphological influence of polypyrrole nanoparticles on the performance of dye–sensitized solar cells. Electrochimica Acta. 2015;**155**:263-271. DOI: 10.1016/j.electacta.2014.12.127

[45] Li CT, Chang HY, Li YY, Huang YJ, Tsai YL, Vittal R, et al. Electrocatalytic zinc composites as the efficient counter electrodes of dye-sensitized solar cells: Study on the electrochemical performances and density functional theory calculations. ACS Applied Materials & Interfaces. 2015;7:28254-28263. DOI: 10.1021/acsami.5b07724

[46] Lin YF, Li CT, Ho KC. A template-free synthesis of the hierarchical hydroxymethyl PEDOT tube-coral array and its application in dye-sensitized solar cells. Journal of Materials Chemistry A. 2016;**4**:384-394. DOI: 10.1039/c5ta06376k

[47] Tsai CH, Huang WC, Hsu YC, Shih CJ, Teng IJ, Yu YH. Poly (o-methoxyaniline) doped with an organic acid as cost-efficient counter electrodes for dye-sensitized solar cells. Electrochimica Acta. 2016;**213**:791-801. DOI: 10.1016/j.electacta.2016.08.012

[48] Li CT, Lin YF, Chiu IT, Ho KC. TCO-free conducting polymers/carbon cloths as the flexible electro-catalytic counter electrodes for dye-sensitized solar cells. Journal of Materials Chemistry A. 2015;**3**:24479-24486. DOI: 10.1039/c5ta06382e

[49] 2012 CoS Acicular Nanorod Arrays for the.pdf

[50] Wu M, Lin X, Wang Y, Wang L, Guo W, Qi D, et al. Economical Pt-free catalysts for counter electrodes of dye-sensitized solar cells. Journal of the American Chemical Society. 2012;**134**:3419-3428. DOI: 10.1021/ja209657v

[51] Zhou H, Shi Y, Dong Q, Wang L, Zhang H, Ma T. High electrocatalytic activity of W18O49 nanowires for cobalt complex and ferrocenium redox mediators. RSC Advances. 2014;**4**:42190-42196. DOI: 10.1039/c4ra07906j

[52] Yang X, Zhou L, Feng A, Tang H, Zhang H, Ding Z, et al. Synthesis of nickel sulfides of different phases for counter electrodes in dye-sensitized solar cells by a solvothermal method with different solvents. Journal of Materials Research. 2014;**29**:935-941. DOI: 10.1557/jmr.2014.74

[53] Sun H, Zhang L, Wang Z-S. Single-crystal $CoSe_2$ nanorods as an efficient electrocatalyst for dye-sensitized solar cells. Journal of Materials Chemistry A. 2014;**2**:16023-16029. DOI: 10.1039/c4ta02238f

[54] Song J, Li GR, Xi K, Lei B, Gao XP, Kumar RV. Enhancement of diffusion kinetics in porous MoN nanorods-based counter electrode in a dye-sensitized solar cell. Journal of Materials Chemistry A. 2014;**2**:10041. DOI: 10.1039/c4ta01342e

[55] Ibrahem MA, Huang W-C, Lan T-W, Boopathi KM, Hsiao Y-C, Chen C-H, et al. Controlled mechanical cleavage of bulk niobium diselenide to nanoscaled sheet, rod, and particle structures for Pt-free dye-sensitized solar cells. Journal of Materials Chemistry A. 2014;**2**:11382-11390. DOI: 10.1039/c4ta01881h

[56] Hsu SH, Li CT, Chien HT, Salunkhe RR, Suzuki N, Yamauchi Y, et al. Platinum-free counter electrode comprised of metal-organic-framework (MOF)-derived cobalt sulfide nanoparticles for efficient dye-sensitized solar cells (DSSCs). Scientific Reports. 2014;**4**:6983. DOI: 10.1038/srep06983

[57] Zhou L, Zhuang Z, Zhao H, Lin M, Zhao D, Mai L. Intricate hollow structures: Controlled synthesis and applications in energy storage and conversion. Advanced Materials. 2017;**29**(20):1602914. DOI: 10.1002/adma.201602914

[58] Singh E, Kim KS, Yeom GY, Nalwa HS. Two-dimensional transition metal dichalcogenide-based counter electrodes for dye-sensitized solar cells. RSC Advances. 2017;7:28234-28290. DOI: 10.1039/c7ra03599c

[59] Dong F, Guo Y, Xu P, Yin X, Li Y, He M. Hydrothermal growth of MoS_2/Co_3S_4 composites as efficient Pt-free counter electrodes for dye-sensitized solar cells. Science China Materials. 2017;**60**:295-303. DOI: 10.1007/s40843-017-9009-8

[60] Chen TY, Huang YJ, Li CT, Kung CW, Vittal R, Ho KC. Metal-organic framework/sulfonated polythiophene on carbon cloth as a flexible counter electrode for dye-sensitized solar cells. Nano Energy. 2017;**32**:19-27. DOI: 10.1016/j.nanoen.2016.12.019

[61] Infant RS, Xu X, Yang W, Yang F, Hou L, Li Y. Highly active and reflective MoS_2 counter electrode for enhancement of photovoltaic efficiency of dye sensitized solar cells. Electrochimica Acta. 2016;**212**:614-620. DOI: 10.1016/j.electacta.2016.07.059

[62] Li CT, Tsai YL, Ho KC. Earth abundant silicon composites as the electrocatalytic counter electrodes for dye-sensitized solar cells. ACS Applied Materials & Interfaces. 2016;**8**:7037-7046. DOI: 10.1021/acsami.5b12423

[63] Fan MS, Lee CP, Li CT, Huang YJ, Vittal R, Ho KC. Nitrogen-doped graphene/molybdenum disulfide composite as the electrocatalytic film for dye-sensitized solar cells. Electrochimica Acta. 2016;**211**:164-172. DOI: 10.1016/j.electacta.2016.06.047

[64] Chiu IT, Li CT, Lee CP, Chen PY, Tseng YH, Vittal R, et al. Nanoclimbing-wall-like $CoSe_2$/carbon composite film for the counter electrode of a highly efficient dye-sensitized solar cell: A study on the morphology control. Nano Energy. 2016;**22**:594-606. DOI: 10.1016/j.nanoen.2016.02.060

[65] Antonelou A, Syrrokostas G, Sygellou L, Leftheriotis G, Dracopoulos V, Yannopoulos SN. Facile, substrate-scale growth of mono- and few-layer homogeneous MoS_2 films on Mo foils with enhanced catalytic activity as counter electrodes in DSSCs. Nanotechnology. 2016;**27**:045404. DOI: 10.1088/0957-4484/27/4/045404

[66] Guo J, Liang S, Shi Y, Hao C, Wang X, Ma T. Transition metal selenides as efficient counter-electrode materials for dye-sensitized solar cells. Physical Chemistry Chemical Physics. 2015;**17**:28985-28992. DOI: 10.1039/c5cp04862a

[67] <2018 The Dye-Sensitized Solar Cells Based on the Interconnected Ternary Cobalt Diindium Sulfide Nanosheet Array Counter Electrode.pdf>

[68] Su L, Xiao Y, Han G, Lin J-Y. One-step hydrothermal synthesis of feather duster-like NiS@MoS_2 with hierarchical array structure for the Pt-free dye-sensitized solar cell. Journal of Nanoparticle Research. 2018;**20**:115. DOI: 10.1007/s11051-018-4223-5

[69] Prasad S, Durai G, Devaraj D, AlSalhi MS, Theerthagiri J, Arunachalam P, et al.

3D nanorhombus nickel nitride as stable and cost-effective counter electrodes for dye-sensitized solar cells and supercapacitor applications. RSC Advances. 2018;**8**:8828-8835. DOI: 10.1039/c8ra00347e

[70] Jiang Y, Qian X, Zhu C, Liu H, Hou L. Nickel cobalt sulfide double-shelled hollow nanospheres as superior bifunctional electrocatalysts for photovoltaics and alkaline hydrogen evolution. ACS Applied Materials & Interfaces. 2018;**10**:9379-9389. DOI: 10.1021/acsami.7b18439

[71] Jiang X, Li H, Li S, Huang S, Zhu C, Hou L. Metal-organic framework-derived Ni–Co alloy@carbon microspheres as high-performance counter electrode catalysts for dye-sensitized solar cells. Chemical Engineering Journal. 2018;**334**:419-431. DOI: 10.1016/j.cej.2017.10.043

[72] Chiu JM, Chen EM, Lee CP, Shown I, Tunuguntla V, Chou JS, et al. Geogrid-inspired nanostructure to reinforce a CuxZnySnzS nanowall electrode for high-stability electrochemical energy conversion devices. Advanced Materials. 2017;7:1602210. DOI: 10.1002/aenm.201602210

[73] Jian SL, Huang YJ, Yeh MH, Ho KC. A zeolitic imidazolate framework-derived ZnSe/N-doped carbon cube hybrid electrocatalyst as the counter electrode for dye-sensitized solar cells. Journal of Materials Chemistry A. 2018;**6**:5107-5118. DOI: 10.1039/c8ta00968f

[74] Huang YJ, Lee CP, Pang HW, Li CT, Fan MS, Vittal R, et al. Microemulsion-controlled synthesis of $CoSe_2/CoSeO_3$ composite crystals for electrocatalysis in dye-sensitized solar cells. Materials Today Energy. 2017;**6**:189-197. DOI: 10.1016/j.mtener.2017.10.004

[75] Huang S, Ma D, Hu Z, He Q, Zai J, Chen D, et al. Synergistically enhanced electrochemical performance of Ni3S4-PtX (X = Fe, Ni) heteronanorods as heterogeneous catalysts in dye-sensitized solar cells. ACS Applied Materials & Interfaces. 2017;**9**:27607-27617. DOI: 10.1021/acsami.7b05418

[76] Huang YJ, Fan MS, Li CT, Lee CP, Chen TY, Vittal R, et al. $MoSe_2$ nanosheet/poly(3,4-ethylenedioxythiophene): Poly(styrenesulfonate) composite film as a Pt-free counter electrode for dye-sensitized solar cells. Electrochimica Acta. 2016;**211**:794-803. DOI: 10.1016/j.electacta.2016.06.086

[77] Li CT, Lee CP, Chiu IT, Vittal R, Huang YJ, Chen TY, et al. Hierarchical TiO1.1Se0.9-wrapped carbon cloth as the TCO-free and Pt-free counter electrode for iodide-based and cobalt-based dye-sensitized solar cells. Journal of Materials Chemistry A. 2017;**5**:14079-14091. DOI: 10.1039/c7ta02474f

[78] Chuang HM, Li CT, Yeh MH, Lee CP, Vittal R, Ho KC. A coral-like film of Ni@NiS with core–shell particles for the counter electrode of an efficient dye-sensitized solar cell. Journal of Materials Chemistry A. 2014;**2**:5816-5824. DOI: 10.1039/c4ta00011k

[79] Lee CT, Peng JD, Li CT, Tsai YL, Vittal R, Ho KC. Ni_3Se_4 hollow architectures as catalytic materials for the counter electrodes of dye-sensitized solar cells. Nano Energy. 2014;**10**:201-211. DOI: 10.1016/j.nanoen.2014.09.017

[80] Shiraz HG, Astaraie FR. Carbonaceous materials as substitutes for conventional dye-sensitized solar cell counter electrodes. Journal of Materials Chemistry A. 2015;**3**: 20849-20862. DOI: 10.1039/c5ta02840j

[81] Batmunkh M, Biggs MJ, Shapter JG. Carbon nanotubes for dye-sensitized solar cells. Small. 2015;**11**:2963-2989. DOI: 10.1002/smll.201403155

[82] Bi H, Cui H, Lin T, Huang F. Graphene wrapped copper–nickel nanospheres on highly conductive graphene film for use as counter electrodes of dye-sensitized solar cells. Carbon. 2015;**91**:153-160. DOI: 10.1016/j.carbon.2015.04.051

[83] Xu Y, Fleischer AS, Feng G. Reinforcement and shape stabilization of phase-change material via graphene oxide aerogel. Carbon. 2017;**114**:334-346. DOI: 10.1016/j.carbon.2016.11.069

[84] Zhang X, Sui Z, Xu B, Yue S, Luo Y, Zhan W, et al. Mechanically strong and highly conductive graphene aerogel and its use as electrodes for electrochemical power sources. Journal of Materials Chemistry. 2011;**21**:6494. DOI: 10.1039/c1jm10239g

[85] Ahmad S, Bessho T, Kessler F, Baranoff E, Frey J, Yi C, et al. A new generation of platinum and iodine free efficient dye-sensitized solar cells. Physical Chemistry Chemical Physics. 2012;**14**:10631-10639. DOI: 10.1039/c2cp41611e

[86] Lee KM, Chen PE, Hsu CU, Huang JS, Ho WS, Chen HH, et al. A high-performance counter electrode based on poly(3,4-alkylenedioxythiophene) for dye-sensitized solar cells. Journal of Power Sources. 2009;**188**:313-318. DOI: 10.1016/j.jpowsour.2008.11.075

[87] Yeh MH, Lee CP, Lin LY, Nien PC, Chen PY, Vittal R, et al. A composite poly(3,3-diethyl-3,4-dihydro-2H-thieno-[3,4-b][1,4]-dioxepine) and Pt film as a counter electrode catalyst in dye-sensitized solar cells. Electrochimica Acta. 2011;**56**:6157-6164. DOI: 10.1016/j.electacta.2011.04.028

Chapter 7

Miniaturized Gas Ionization Sensor Based on Field Enhancement Properties of Silicon Nanostructures

Parsoua Abedini Sohi and Mojtaba Kahrizi

Abstract

According to principle of the operation, gas field ionization sensors are classified as transduction-based gas sensors. These sensors identify the unknown gases based on their unique ionization properties such as breakdown voltage or tunneling current. Appling 1D nanostructure in gas ionization sensors would enhance the local electric field at the tip of the structures. The average field enhancement coefficient (β_{tol}), considering constructive/destructive interferences of the local electric field of thousands of nanowires in the whole structure, is desired to optimize the design and structure of the gas sensors. Using chemical/electrochemical techniques silicon nanowires were grown on one of the electrodes of the gas sensor. Mechanism of the nanowires formation was modeled and simulated using COMSOL multiphysics simulation tool prior to their fabrication. A gas field ionization tunneling sensor, was designed, fabricated, and tested successfully for several gases like N_2, He, and Ar. Estimated β_{tol} of the sensor showed that the electric field strength inside the sensor is 3750 times greater than a planar parallel-plate sensor causing to reduce the breakdown voltages from several thousand volts to the range of 60–70 V for various gases.

Keywords: gas ionization sensor, tunneling current, field enhancement factor, COMSOL, silicon nanowires

1. Introduction

There are several natural and artificial chemical species in the air, some of which are toxic and combustible gases, which can be considered as potential hazard to the health [1–3]. Gas sensors helps to prevent these hazards and play an important role in places with the risk of potentially harmful substances specifically in industrial processes and manufacturing plants.

Depending on the principal of the operation, gas sensor devices can be classified into two distinct categories: chemical gas sensors and physical gas sensors [3]. Merely, considering transduction mechanism, chemical gas sensors are based on chemical reaction between gas and sensing materials, resulting in change of conductivity of the detector material. In this method, gases in different combinations could produce the same net-change in conductance and as a result, distinguishing between a gas and its mixtures is impossible. These types of sensors are very sensitive to changes in moisture, temperature and gas pressure. Their other drawback is that chemical reactions could cause irreversible changes in detector materials [3]. Physical type sensors have overcome the disadvantages of chemical type sensors. There are several physical type gas sensors, according to the mechanism of operation, including surface plasmon resonance (SPR) based [4] gas sensors, fiber optic based gas sensors [5–8] and gas ionization sensors [9].

Surface plasmon resonance (SPR) is the resonant oscillations of surface electrons, which are stimulated by incident illumination at the interface between a metal and dielectric [10, 11]. SPR is very sensitive to the refractive index of the medium close to the metal film. The resonance spectral response of the SPR will change when the conditions of the medium are changed, which can reflect certain properties of the system. Kretschmann geometry (prism coupler) is widely used to study SPR. In this configuration, optical wave is totally reflected at prism-metal interface [12]. Evanescent field wave may penetrate the metal layer and excite surface plasmon at the metal-dielectric boundary. As excitation of surface plasmon significantly reduces the intensity of reflected light, reflectivity of the sensor as a function of either wavelength or incident angle is considered as the sensor response [12].

Agbor et al. [13] reported the SPR gas sensing measurements in Kretschmann configuration using nickel/silver coated glass microscope slides. According to their results, the SPR curves were influenced by 50 ppm of NO_2 and H_2S at room temperature. Maharana et al. [14] reported a numerical study on a high performance SPR sensor based on graphene coated silver on wide range of refractive indices of gases. Graphene is widely used in SPR based gaseous detection systems, as its refractive index is highly sensitive to the absorbed gas molecules. Furthermore, graphene is robust against the oxidation and the layer of graphene in SPR sensors (in presence of noble metals) prevents oxidation of the silver layer. Nooke et al. [15] studied the SPR gas sensing measurements in Kretschmann configuration using gold (Au) coated glass for combustible, toxic and greenhouse gases. They also reported that the gas detection limit is related to the rate of gas adsorption, which is defined by polarizability of the gases.

SPR-based fiber optic sensors are designed by replacing the cladding with a thin layer (in nm range) of metal. In these sensors it is hard to reach the sensitivity similar to Kretschmann SPR configuration due to complexity in controlling the incidence angle of light, impossibility to control the wave polarization and an excessive number of reflections. In these sensors the spatial-frequency bandwidth of their angular spectrum is wider in comparison with other types of SPR sensors [16, 17]. However, some noticeable advantages like low cost, flexibility, real-time monitoring, compatibility with human tissue and blood vessels, remote sensing, small sample volume, reusability, and simple structure have made the SPR fiber optic approach very attractive.

Gas sensing application of the fiber optic sensor was developed in 1980 [1] and the sensing measurements are essentially based on changing the features of transmitted light along the fiber. Transmitted light can be modified in response to external medium properties. According to the principal of the operation, fiber optic gas sensor devices can be classified into two distinct categories: extrinsic and intrinsic [1, 6]. In extrinsic fiber optic gas sensors, light exits the fiber and interacts with the medium before continuing propagation inside the core again. In these sensors, light propagates through the input fiber optic toward a microcell containing the unknown gas. The output signal is guided to a spectrometer using the output fiber optic, which is accurately aligned with the input one. This provides the unknown gas detection by comparing the input interrogating wavelength and the absorption spectrum of the gas [6]. This technique can be only used for the gases which spectral absorption is in the range of telecommunication window so the fiber can be successfully employed. Stewart et al. [18] reported a design of fiber optic methane sensor using a microcell and DFB laser source. The theoretical modeling of the designed sensor was in a good agreement with experimental results and both showed a single absorption line for the methane gas.

Shemshad et al. [19, 20] investigated the absorption band of methane and stated that absorbance spectrum of methane is between 1620 and 1700 nm which is suitable for fiber optic detection sensors. As methane is released during coal extractions, they have also studied the cross sensitivity of methane with other gases emitted from mine. Their results showed that the absorbance spectrum of methane does not interfere with other gases in the mine.

In intrinsic fiber optic gas sensors, light propagates inside the core continuously without any external interaction. When a light is propagating through fiber optics, at the core-cladding boundary it undergoes total internal reflection (TIR). During each TIR it penetrates into the cladding region, which is known as evanescent wave. The amplitude of the evanescent wave decays exponentially in the cladding region. So the cladding (with lower refractive index) absorbs a small portion of propagating light energy. This process is known as attenuated total reflection (ATR). If an absorbing chemical or testing sample is present with the evanescent field region, the propagating light will be attenuated (as the reflection coefficient is less than unity) as it travels along the fiber. Since the energy levels associated with an atom or molecules are unique, the absorption spectrum serves as a "fingerprint" identification of the chemical species [1]. To increase the sensitivity of such sensors the cladding can be manufactured to be sensitive to specific organic vapors [21] or an unclad fiber can be coated with sensitive coatings [22]. Another alternative way is diminishing the cladding thickness, which results in a more fragile but more sensitive sensor [6, 20].

Gas field ionization sensors identify unknown gases by their unique ionization characteristics [23]. Calibration of these devices is based on fingerprinting breakdown voltage of the target gases. The gas sensor is made of two parallel planar electrodes. The two plates are separated by pieces of insulating films such a way that there is enough opening between the two electrodes to allow the gas flow between them. Applied voltage across the device is swept and insulation-to-conduction transformation of the gas (known as breakdown), using I-V characteristics of electrical discharge, is recorded [23]. This technique has improved selectivity property of the gas sensors, as it is approved that at a constant pressure and temperature, each gas has unique breakdown voltage (or breakdown electric field) [9]. An applied voltage required to generate breakdown electric field (E_{app}), depends on the separation gap between the plates and typically is over several hundred volts [24–26]. As applying this voltage is not practical, to scale down the breakdown voltage of the gases, 1D nanostructures could be applied as one of the electrodes. Enhanced local electric field (E_{loc}) is created at the tip of the nanostructures due to non-uniform distribution of charged carriers [27, 28]. Field enhancement factor (β) characterizes the level of influence of the 1D nanostructures onto the electrostatic field distribution and can be defined as the field gain coefficient ($\beta = E_{loc}/E_{app}$) [27].

In this work we have explored an example of gas ionization sensor that is fabricated based on p-type silicon (Si) nanostructures. The ionization characteristics of several gases are reported. The effect of applying these structures on the field enhancement factor of the sensor, compared to a parallel plate system, is described.

2. Geometrical field enhancement

Nanowires amplify E_{loc} regardless of their bias direction. However depending on the type of the materials of the nanowires some promote field emission, and others will enhance field ionization tunneling phenomenon.

In field emission based gas sensors, metallic or highly doped n-type semiconductor nanostructures are applied as the cathode (negatively biased), in which enhanced local electric field induces the emission of the electrons from the tip of the nanostructures [9]. In this phenomenon, as the name suggested, under a strong electric field electrons will overcome the deformed potential barrier of the vacuum and will be emitted from the surface of the cathode [29]. The diagram shown in **Figure 1a** qualitatively illustrates the electron emission from a metallic cathode. The shape and width of the metal-surface potential barrier is influenced by the strength of the applied electric field [30]. When the barrier is narrow enough electrons can escape through the barrier, thereby forming a field electron emission. In a gas sensor, field emission dominates the breakdown mechanism of the gases and reduces the breakdown voltage significantly. This phenomenon occurs as field emitted electrons contribute in impact ionization of gas atoms. There are several materials reported for field emission applications. Among these materials carbon nanotubes have been actively studied as field emitters [31]. The use of multi-walled carbon nanotubes is frequently reported in field emission gas ionization sensors, which reduced the breakdown voltage of the gases compared to planar parallel plate [32–34]. Metallic nanowires such as gold and silver have also shown field emission properties and are employed as the cathode in gas ionization sensors [35, 36].

In field ionization tunneling sensors, p-type semiconductor nanostructures are applied as the anode (positively biased). Due to enhanced electric field at the tip of the nanostructures, valence electrons of gas atoms can escape from the atom by tunneling through the nucleus potential barrier into available energy states of the p-type nanostructures (as shown in **Figure 1b**). In these sensors tunneling currents can be used as the calibrating data as it occurs at lower voltages compared to complete breakdown [36, 37]. However in some cases (depending on the separation gap distance, gas pressure and the material of the cathode) bombardment of the cathode by the released positive ions may result in secondary emissions, which can contribute to either quasi or complete breakdown [38].

There are several analytical studies reported in literature to calculate field enhancement factor (β) of the nanostructures [39, 40]. These studies mathematically predicted the field enhancement factor of a single protrusion; however, the interpretation of the average field gain coefficient (β_{tol}), considering constructive/destructive interferences of the local electric field of thousands of nanowires in the whole structure, is desired to optimize the design and structure of the gas sensors.

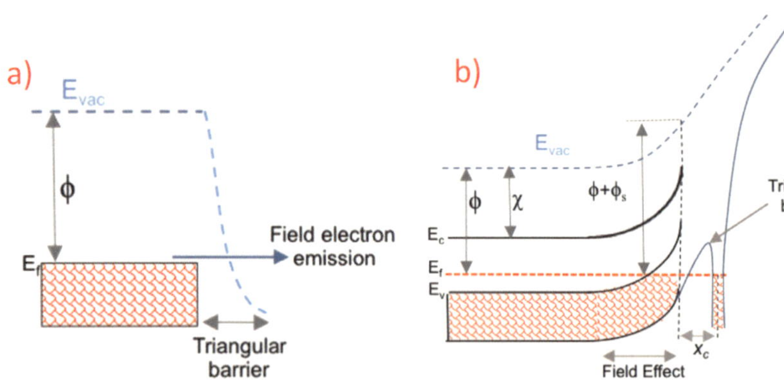

Figure 1.
(a) Field electron emission mechanism from metallic cathode. At large electric fields the potential barrier is bended, which allows electrons escape through the barrier toward the vacuum, (b) valence electrons of gas atom, tunnels through the potential barrier of the nucleus into the available energy states of p-type anode (part b is reproduced from Abedini Sohi & Kahrizi [38]).

In field emission based gas sensors, Fowler Nordheim (FN) theory of electron emission from a metallic or semi-metallic roughly predicts β_{tol} based on the field induced emission current and can be experimentally calculated by plotting ln (J/V^2) against (1/V) where J is the emission current density and V is the applied voltage [41, 42].

In the second approach, this factor can be estimated practically using the slope of the gas discharge graph (I-V characteristics) in the ohmic region of the curves (for field emission gas sensors as well as field ionization tunneling sensors). **Figure 2** shows different regions of discharge characteristic of gases in a uniform electric field generated between planer parallel plates. Any gas contains free-floating negative electrons and positive ions due to the ionization of the gas atoms by cosmic radiation. In region I of the discharge curve, when the voltage is applied to the electrodes, the traveling radiation-generated charged particles produce current as electrons migrate to the anode and the ions move toward the cathode. The current in this region shows strong electric field dependence and increases as the applied voltage is increased. In this region the discharge current density is expressed as:

$$J_{GIS} = \sigma_{Gas} \times E_{app} = (n_e \mu_e + n_i \mu_i) \times e \times E_{app} \tag{1}$$

where σ_{Gas} is the gas conductivity, e the electron charge, E_{app} the applied electric field, n_e and n_i the electron and ion concentrations respectively, and μ_e, and μ_i are the electron and ion mobility respectively.

In saturation region (II), the current reaches its saturated value which means that all radiation-generated particles are attracting to the electrodes. At the last phase the electric field strength would be enough to accelerate the electrons, resulting ionizing collisions. At breakdown voltage (V_b) the discharge current is self-sustainable and is maintained due to ionizing collisions without any external ionization source.

According to Eq. (1) in the ohmic region, the traveling radiation-generated charged particles produce current proportional to applied field. As in this region there is no ionization-induced current, the current density exclusively depends on the strength of the electric field. The current density of nanowire based gas ionization sensor (J_{GIS}) due to enhanced electric field can be expressed as

$$J_{GIS} = \sigma_{Gas} \times E_{eff} = (n_e \mu_e + n_i \mu_i) \times e \times \beta_{tol} \times E_{app} \tag{2}$$

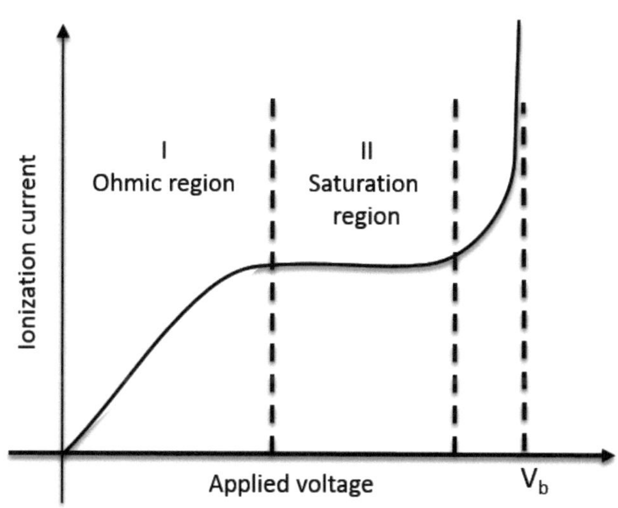

Figure 2.
I-V characteristics of gas discharge in a uniform electric field.

where E_{eff} is the enhanced electric field and β_{tol} is average field gain coefficient.

By comparing Eqs. (1) and (2) and considering a constant σ_{Gas}, β_{tol} of gas ionization sensors can be estimated by dividing the slopes of I-V characteristics of the device with a parallel-plates in the ohmic region.

$$\beta_{tol} = \frac{Slope_{GIS}}{Slope_{PPL}} \tag{3}$$

where $Slope_{GIS}$ is the slope of I-V discharge graph of the nanowire based GIS and $Slope_{PPL}$ is the slope of the I-V discharge graph of the parallel plate in their ohmic regions.

3. Fabrication of silicon nanostructures

Silicon nanostructures are fabricated using chemical/electrochemical technique [43]. Samples of p-type <100> silicon wafers (380 ± 10 μm thickness with resistivity about 5–10 Ω cm from Silicon Material Inc.) were cut in 1 × 1 cm^2 pieces and cleaned using RCA technique. In the first step of etching, samples were textured by pyramidal structures through anisotropic etching in tetramethylammonium hydroxide (TMAH) based solution. The solution for anisotropic etching, made of equal amount of TMAH and isopropyl alcohol (IPA) (5 wt% each), was used to etch the samples for 20 minutes. The temperature of the solution was kept constant at 90°C using an oil bath system. A condenser covered the etchant container in order

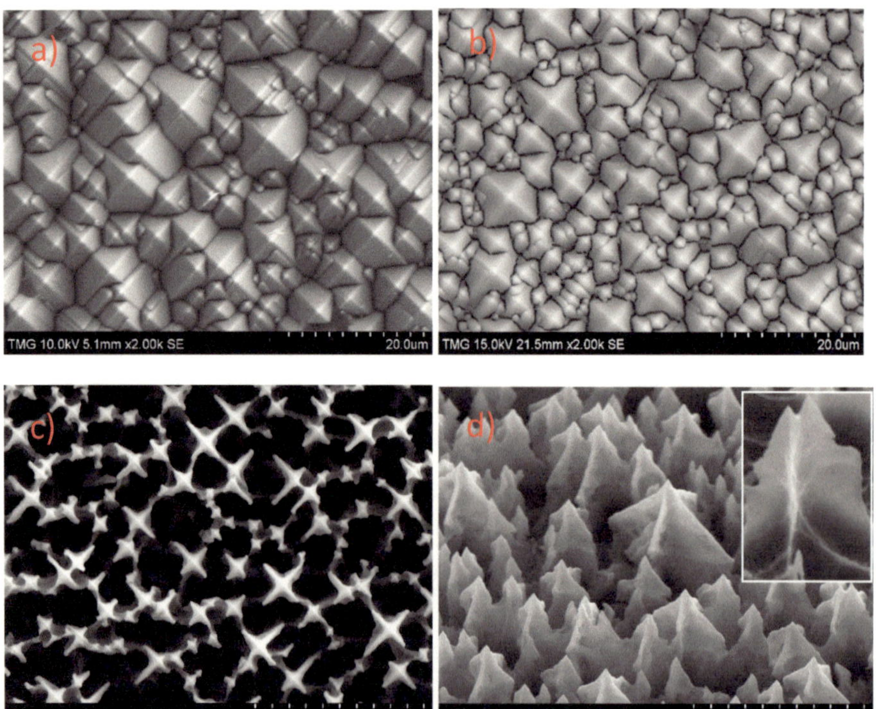

Figure 3.
Experimental results of the evolution of the nanostructures. (a) Anisotropic etching of <100> p-type silicon at 90°C for 20 minutes in 5 wt% TMAH and 5 wt% IPA resulted in formation of pyramids that covers the surface completely. (b) Top view of the electrochemically etched textured silicon after 10 minutes of etching. (c) Top view of the electrochemically etched silicon after 70 minutes. (d) Top view of the electrochemically etched silicon after 70 minutes. (d-inset) Tilted view of the final nanostructures illustrates the existence of residual walls around the structures with the thickness of 100 nm.

to keep the etchant concentration constant throughout the experiment. **Figure 3a** shows the SEM image of textured Si sample containing square-based pyramidal structures after anisotropic etching step. As it is shown in the figure, the average tip-to-tip separation of the large pyramids is more than 5 μm. The samples were subsequently exposed to electrochemical isotropic etching. In this step the textured silicon samples, are placed in a two-electrode cell of anodic etching. The etching was done in an electrolyte consisting of 1:3 hydrofluoric acid and ethanol for 70 minutes under a 15 mA/cm^2 applied anodic current density. Porous silicon starts to form by electrochemical etching, all around the pyramids. **Figure 3b** shows the SEM image of electrochemically etched sample after 10 minutes of etching. As the etching continues, the pyramid faces are etched away, while lateral edges remain unetched. **Figure 3c** and **d** shows the top view and tilted view SEM image of the sample after 70 minutes isotropic etching. To narrow down the lateral edges, samples are subjected to another anisotropic etching in a weak etchant solution (1 wt% of TMAH and 1 wt% of IPA) for 30 seconds. The final structures are shown in the inset of **Figure 3d**. They are arrow shape structures with remaining lateral edges with the thickness about 100 nm.

4. Formation mechanism of the nanostructures

Electrochemical dissolution of Si, in the second step of etching, is highly influenced by hole current density at the interface of Si/electrolyte. Hole drift current density in semiconductors can be found by,

$$J_p = e \times p \times \mu_p \times E\left(\frac{A}{cm^2}\right) \tag{4}$$

Where e is electric charge (1.6×10^{-19} coulomb), p is the concentration of holes/cm^3, μ_p is the hole mobility (cm^2/V s) and E is the applied electric field (V/cm).

Eq. 4 confirms the dependence of the etching rate with the electric field strength. According to **Figure 3b** the pyramid perimeters are experiencing a higher etching rate at the beginning of the etching process. This can be attributed to maximum electric field intensity at those regions.

Due to uneven textured Si surfaces, the electric field is not uniform at the interface. As a result, the hole drift current density is not uniform over the surface of the samples and different areas are experiencing different dissolution rates during the anodic etching.

The electric field intensity generated at textured Si (anode) is modeled and simulated utilizing COMSOL, electrostatics module. To study the evolution mechanism we have considered the 3D schematic geometry and initial electric field distribution shown in **Figure 4** when voltage is applied to the backside of the textured Si. Simulation results shows that the electric field is not uniform and the lowest belongs to tips and lateral edges of the pyramids, and highest is for pyramid perimeters.

To present the development of the structures, COMSOL deformed geometry (dg) physics interface is applied to all domains. A prescribed mesh velocity is assigned to the boundary between Si and electrolyte (pyramidal texture) and a prescribed mesh displacement is assigned to boundaries all around the block. Moving mesh velocity is defined proportional to electric field.

Figure 5 clearly shows that the etching and developing porous silicon start at the edges of the pyramids bases where the electric field has the maximum value.

As the etching extends, pyramid faces are etched away while lateral edges remain as connecting walls between the pyramids. **Figure 6** clearly illustrates the residual walls between the structures.

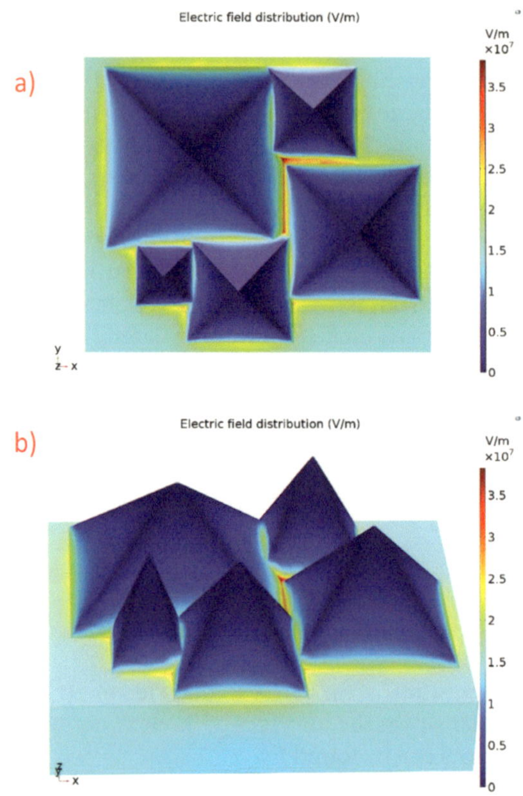

Figure 4.
Distribution of electric field in the textured Si surface is simulated by COMSOL electrostatics module. (a) Top view and (b) tilted view of the 3D structures shows that the electric field strength is not uniform and the lowest belongs to vertex and sides (shown by dark blue) and highest is for pyramid perimeters (shown in yellow/red).

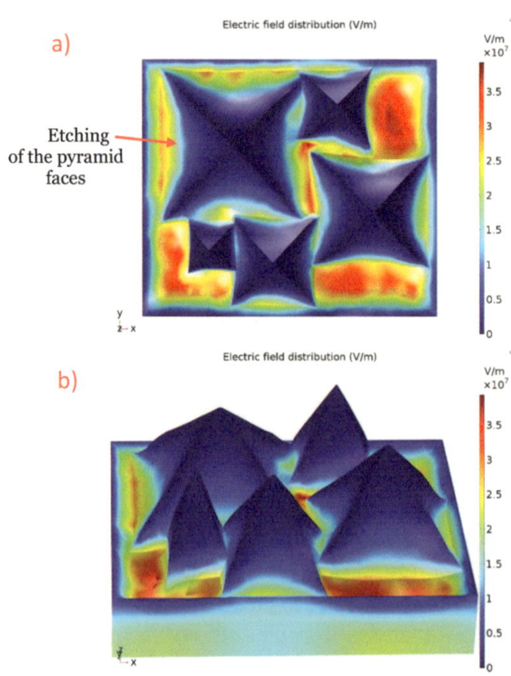

Figure 5.
(a) Top view and (b) tilted view of COMSOL deformed geometry. The figures show that as the etching starts, pyramid perimeters are experiencing higher etching rates due to higher electric filed strength.

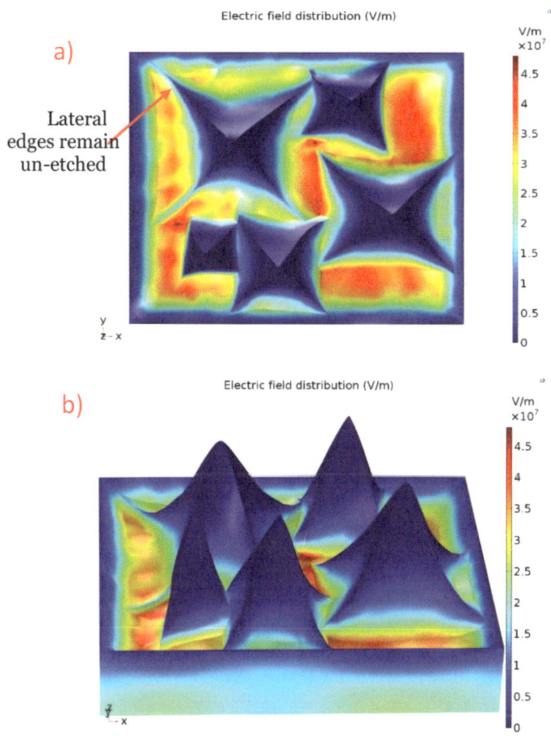

Figure 6.
COMSOL deformed geometry shows that the extended etching results in residual walls around the structures, while pyramid faces are etched away. (a) Top view, (b) tilted view.

5. Device fabrication and sensor setup

Fabricated Si nanostructures were incorporated as the anode in gas ionization cell. The cathode is a piece of p-type Si cleaned with RCA method and coated with a 5 μm aluminum (Al) layer using thermal evaporation technique. The electrodes are separated, using an insulating thin film (double sided adhesive tape), by a narrow gap but wide enough to allow the flow of the gases through the cell. **Figure 7** shows the schematic illustration of the gas ionization cell.

The device was placed in the gas chamber and vacuumed to 10^{-5} Torr prior to introduce each gas to the chamber. The electrodes are connected to two source measure units (SMUs) of a HP4155 semiconductor parameter analyzer and I-V characteristics of the device was conducted by sweeping the voltage of the anode

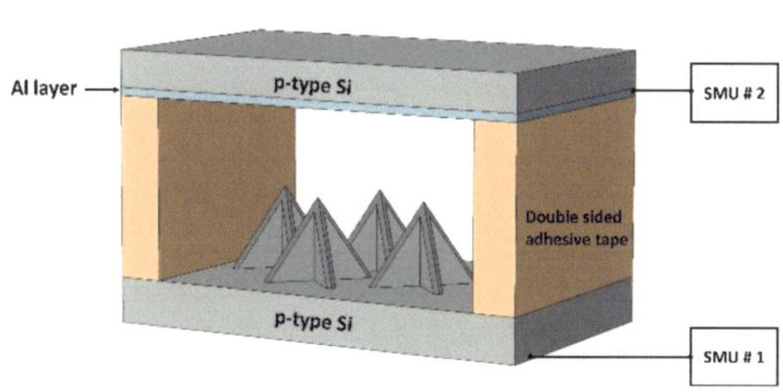

Figure 7.
Schematic of Si nanowires-based GITS. Fabricated sample is applied as the anode and it attached to SMU #1 of the parameter analyzer. An Al coated Si is applied as the cathode. Electrodes are separated by two pieces of double sided adhesive tape. The other sides are left open to facilitate the gas flow through the sensor.

from 0 to 100 V, with steps of 1 V and 1 second sweep delay. The sensor was tested for oxygen (O_2), argon (Ar), Nitrogen (N_2) and helium (He) at low pressures.

6. Results and discussions

6.1 Estimation of β_{tol}

To calculate the total field gain coefficient (β_{tol}) based on Eq. (3), I-V characteristics of a PPL sensor is compared to that of the GITS. The parallel-plates sensor consists of two plates, one perfectly smooth p-type Si wafer as the anode and an aluminum coated Si wafer as the cathode, separated by 100 μm gap. GITS sensor is fabricated as explained in Section 5 with the same separation gap of 100 μm between the electrodes.

Ohmic regions of both PPL sensor and fabricated GITS, under vacuumed air (10^{-5} Torr), are shown in **Figure 8**. The total enhancement factor of the device is calculated using Eq. 3. Accordingly, the enhancement factor of the device is equal to:

$$\beta_{tol} = \frac{Slope_{GIS}}{Slope_{PPL}} \approx \frac{3 \times 10^{-5}}{8 \times 10^{-9}} = 3,750 \quad (5)$$

Theoretically, no tunneling occurs at electric fields less than 10^9 V/m. The electric filed strength in the parallel plate sensor is in the range of 10^5–10^6 V/m according to applied voltage (100 V is the maximum voltage of HP4155-SMU). This value should at least increases by 1000 times to make the tunneling possible. Our estimated value of β_{tol}, indicates that the geometrical field enhancement of fabricated GITS induces the required electric field for the field ionization tunneling.

6.2 Characterization of the gas sensor

Field ionization tunneling tests were performed at 10^{-2} Torr for several gases, while separation gap between the electrodes was set at 200 μm. A 20 mA current compliance is applied to both SMUs in order to protect nanowires from burning.

Figure 9 shows the same trend for I-V characteristics of all the gases. Below 20 V the gas discharge is in its ohmic region, in which current is low and is due to the movement of radiation-generated charged particles. At the voltages above the ohmic region, contrary to ionization discharge curve of a PPL sensor (shown in

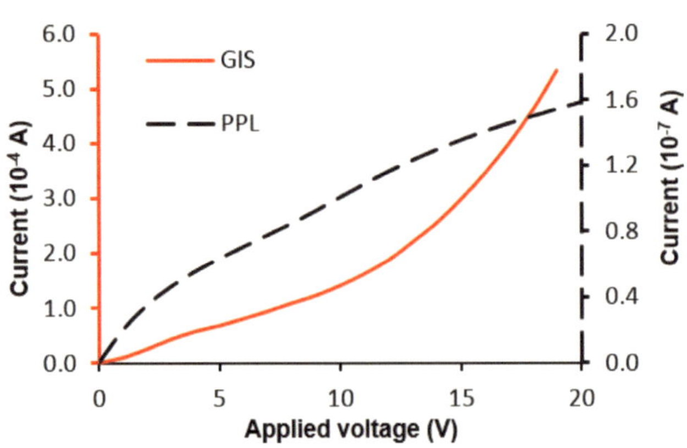

Figure 8.
Ohmic regions of the PPL sensor and the fabricated GITS at 10^{-5} Torr vacuumed air. The slopes of the curves are used to calculate the enhancement factor of the system.

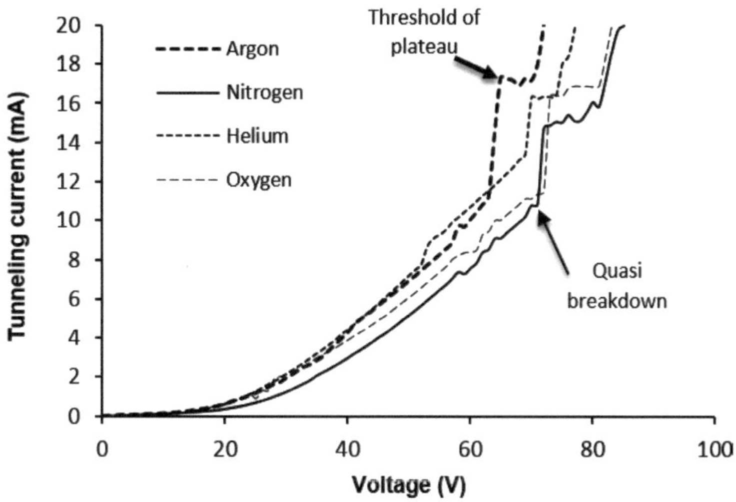

Figure 9.
Room temperature I-V characteristics of Ar, N_2, He and O_2 at 10^{-2} Torr show distinctive tunneling properties of the gases. The transition from ohmic region to tunneling region can be seen at 20 V, where the tunneling current increased steeply (reproduced from Abedini Sohi & Kahrizi [38]).

Figure 2) which reaches to a saturated current, the discharge current of fabricated GITS increases sharply as the applied voltage is increased. In this region the electric field is higher than ionization threshold and the valence electrons of the gases are able to tunnel through the potential barrier of the gas atom, into the unoccupied energy states of the anode electrode.

Each gas reaches to a quasi-breakdown (V_{qbr}) at a specific voltage. At V_{qbr}, an abrupt rise in the tunneling current is due to the cathode bombardment by positive ions, which results in gas amplification due to secondary emissions of the Al layer. Large number of positive ions is generated during the gas amplification and move toward the cathode to get neutralized. These positive ions create an internal electric field in opposite direction to the applied electric field. Consequently the total electric field is reduced, which reduces the rate of ionization and as a result the tunneling current reaches a plateau. After neutralization of the accumulated positive ions, the tunneling current is increased again until reaching current compliance.

The V_{qbr} and threshold of plateau can be used as calibrating data to distinguish the gases.

Figure 10 represents the room temperature I-V characteristics of the fabricated GITS for He (**Figure 10a**) and Ar (**Figure 10b**) at different pressures. According to the results, no correlation between the tunneling current and pressure can be addressed. However, the V_{qbr} for the both gases is raised as the pressure is increased. This increase in V_{qbr} can be explained by reduced mean free path (the average distance traveled by a particle to make successive ionizing collisions). As the mean free path is reduced in higher pressures, the electrons acquire enough energy to create impact ionizations at higher electric fields.

The experiments are extended to test the effect of separation gap between the electrodes on I –V characteristics of the sensor. For this reason, the same device with 50, 100 and 200 μm separation gaps was tested for Ar at 10^{-2} Torr. As it is shown in **Figure 11**, by reducing the separation gap to 50 μm, the sensor showed a complete breakdown at 50 V due to the high electric field strength. At 100 μm a complete breakdown was observed at 60 V. Lower electric field in the devices with 200 μm separation gap, resulted in quasi breakdown and a threshold of plateau that is explained earlier.

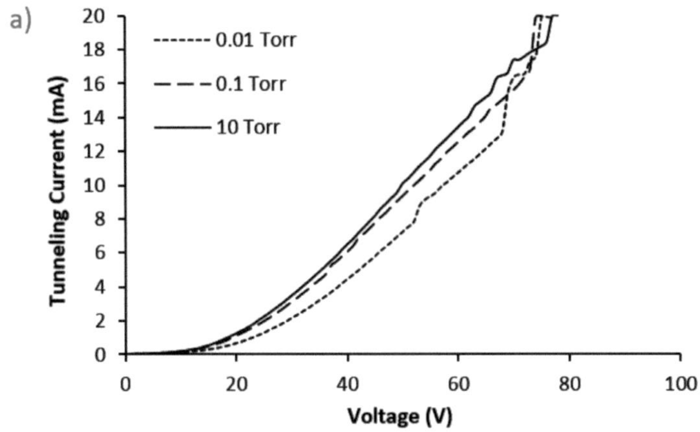

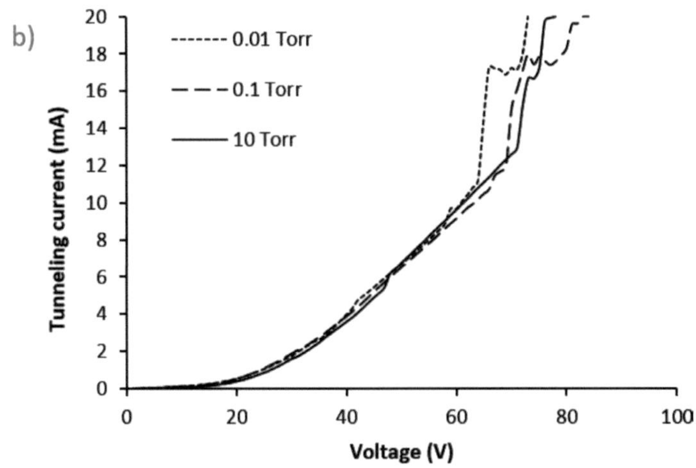

Figure 10.
Room temperature I – V characteristics of He (a) and Ar (b) at a wide range of pressures (0.01–10 Torr) (part (b) is reproduced from Abedini Sohi & Kahrizi [38]).

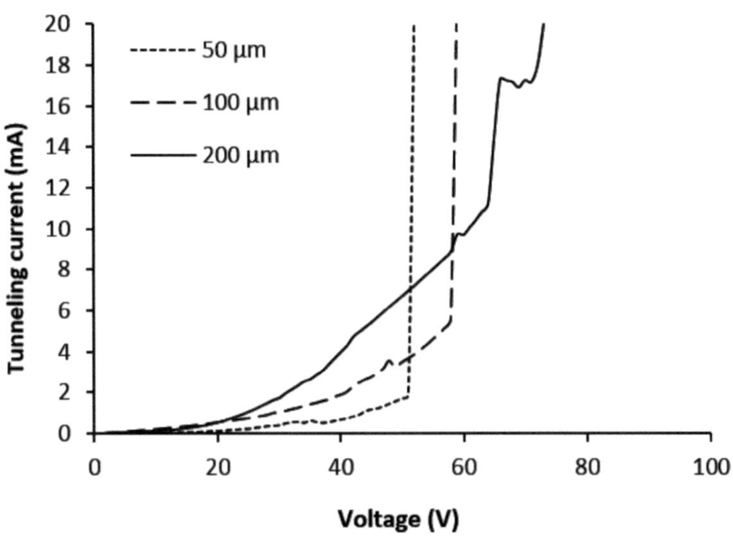

Figure 11.
Fabricated GITS is tested for Ar at 10^{-2} Torr for different separation gaps. By increasing the separation gap between the electrodes, the transition from complete breakdown to quasi breakdown is observed.

7. Conclusion

A novel fabrication technique, based on consecutive chemical and electrochemical etching techniques, is used to fabricate Si nanostructures. Si surface is textured by pyramidal hillocks through anisotropic etching in TMAH based solution. Electrochemical etching (anodic etching) of the textured Si was carried out in a HF-based solution in an electrochemical cell. Non-uniform distribution of the electric field induces different level of etching rate over the anode, which results in formation of the arrow shape structures. Mechanism of the developed structures is investigated by modeling and simulation by COMSOL multiphysics. Fabricated structures were applied as the anode in GITS. The total field enhancement coefficient (β_{tol}) of the GITS is estimated based on the ohmic region of the gas discharge characteristics, as compared to a parallel-plates sensor. Field penetration and band bending at the surface of p-type nanostructures lead to tunneling current in the range of mA in low voltages and as a result, the fabricated Si nanostructure based GITS showed the capability to distinguish the unknown gases as well as the gas pressure.

Acknowledgements

This work was partially supported by the Natural Sciences and Engineering Research Council of Canada (NSERC) and by the Gina Cody school of Engineering and Computer Science at Concordia University.

Conflict of interest

Parsou Abedini Sohi and Mojtaba Kahrizi declare that this article does not contain any conflict of interest.

Author details

Parsoua Abedini Sohi and Mojtaba Kahrizi*
Concordia University, Montreal, QC, Canada

*Address all correspondence to: mojtaba.kahrizi@concordia.ca

References

[1] Mandelis A, Christofides C. Physics, Chemistry and Technology of Solid State Gas Sensor Devices. New York: Wiely; 1993

[2] Korotcenkov G. Handbook of Gas Sensor Materials. New York, Heidelberg Dordrecht London: Springer; Vol. 1 and 2; 2014

[3] Sadeghian RB. A Field Effect Gas Sensor Based on Self-Standing Nanowire Array. Montreal, Canada: Concordia University; 2007

[4] Liedberg B, Nylander C, Lunstrom I. Surface plasmon resonance for gas detection and biosensing. Sensors and Actuators. 1983;**4**:299-304

[5] Sharma AK, Pandey AK, Kaur B. A review of advancements (2007-2017) in plasmonics-based optical fiber sensors. Optical Fiber Technology. 2018;**43**:20-34

[6] Elosua C, Matias IR, Bariain C, Arregui FJ. Volatile organic compound optical fiber sensors: A review. Sensors. 2006;**6**(11):1440-1465

[7] Gupta BD, Verma RK. Surface plasmon resonance-based fiber optic sensors: Principle, probe designs, and some applications. Journal of Sensors. 2009;**2009**:12

[8] Hodgkinson J, Tatam RP. Optical gas sensing: A review. Measurement Science and Technology. 2013;**24**(1):1-59

[9] Howaston AM. An Introduction to Gas Discharges. 2nd ed. Oxford: Pergamon Press; 1976

[10] Liang J. Surface plasmon resonance: An introduction to a surface spectroscopy technique. Journal of Chemical Education. 2011;**87**(7):742-746

[11] Homola J. Surface Plasmon Resonance Based Sensors. Springer; 2006

[12] Manera MG, Rella R. Improved gas sensing performances in SPR sensors by transducers activation. Sensors and Actuators B: Chemical. 2013;**179**:175-186

[13] Agbor NB, Cresswell JP, Petty MC, Monkman AP. An optical gas sensor based on polyaniline Langmuir-Blodgett films. Sensors and Actuators B: Chemical. 1997;**41**(1-3):137-141

[14] Maharana PK, Jha R, Padhy P. On the electric field enhancement and performance of SPR gas sensor based on graphene for visible and near infrared. Sensors and Actuators B: Chemical. 2015;**207**(Part A):117-122

[15] Nooke A et al. On the application of gold based SPR sensors for the detection of hazardous gases. Sensors and Actuators B: Chemical. 2010;**149**(1):194-198

[16] Roh S, Chung T, Lee B. Overview of the characteristics of micro- and nano-structured surface plasmon resonance sensors. Sensors. 2011;**11**(2):1565-1588

[17] Tosi D, Poeggel S, Iordachita I, Schena E. Fiber optic sensors for biomedical applications. In: Opto-Mechanical Fiber Opt Sensors. 2018. pp. 301-333

[18] Stewart G, Tandy C, Moodie D, Morante MA, Dong F. Design of a fibre optic multi-point sensor for gas detection. Sensors and Actuators B: Chemical. 1998;**51**(1-3):227-232

[19] Shemshad J. Design of a fibre optic sequential multipoint sensor for methane detection using a single tunable diode laser near 1666 nm. Sensors and Actuators B: Chemical. 2013;**186**:466-477

[20] Shemshad J, Aminossadati SM, Kizil MS. A review of developments in near infrared methane detection based on tunable diode laser.

Sensors and Actuators B: Chemical. 2012;**171-172**:77-92

[21] Schwotzer G, Latka I, Lehmann H, Willsch R. Optical sensing of hydrocarbons in air or in water using UV absorption in the evanescent field of fibers. Sensors and Actuators B. 1997;**39**:38-39

[22] Mishra SK, Tripathi SN, Choudhary V, Gupta BD. SPR based fibre optic ammonia gas sensor utilizing nanocomposite film of PMMA/reduced graphene oxide prepared by in situ polymerization. Sensors and Actuators B: Chemical. 2014;**199**:190-200

[23] Hall F, Hill B. Plasma Gas Disch, Vol. 92000. pp. 1-11

[24] Boyle WS, Kisliuk P. Departure from Paschen's law of breakdown in gases. Physics Review. 1955;**97**(2):255-259

[25] Lisovskii VA, Yakovin SD. A modified Paschen law for the initiation of a dc glow discharge in inert gases. Technical Physics. 2000;**45**(6):727-731

[26] Radmilović-Radjenović M, Radjenović B. An analytical relation describing the dramatic reduction of the breakdown voltage for the microgap devices. Europhysics Letters. 2008;**83**(2):25001

[27] Chivu N, Kahrizi M. Design, Modeling and Simulation of a Miniaturized Gas Ionization Sensor: Optimization of the Structure and Operation; 2012. pp. 178-183

[28] Chivu N, Kahrizi M. Modeling and simulation of a miniaturized gas ionization sensor: Detection of greenhouse gases. Sensors & Transducers. 2013;**153**(6):105-110

[29] Jensen KL, Shiffler DA, Harris JR, Rittersdorf IM, Petillo JJ. 2D/3D image charge for modeling field emission. Journal of Vacuum Science & Technology, B: Nanotechnology & Microelectronics: Materials, Processing, Measurement, & Phenomena. 2017;**35**(2):02C101

[30] Yilmazoglu O et al. Electron field emission from nanostructured semiconductors under photo illumination. Turkish Journal of Physics. 2014;**38**(3):543-562

[31] Saito Y, Uemura S. Field emission from carbon nanotubes and its application to electron sources. Carbon. 2000;**38**(2):169-182

[32] Modi A, Koratkar N, Lass E, Wei B, Ajayan PM. Miniturized gas ionization sensors using carbon nanotubes. Nature. 2003;**424**:171-174

[33] Yeow JTW, Wang Y. A review of carbon nanotubes-based gas sensors. Journal of Sensors. 2009;**2009**:24

[34] Kim SJ. Gas sensors based on Paschen's law using carbon nanotubes as electron emitters. Journal of Physics D: Applied Physics. 2006;**39**(14):3026-3029

[35] Sadeghian RB, Kahrizi M. A novel miniature gas ionization sensor based on freestanding gold nanowires. Sensors and Actuators, A: Physical. 2007;**137**(2):248-255

[36] Azmoodeh N, Chivu N, Sadeghian RB, Kahrizi M. A silver nanowire based gas ionization sensor EUROCON 2009. In: IEEE EUROCON 2009. 2009. pp. 1231-1235

[37] Sadeghian RB, Kahrizi M. A novel gas sensor based on tunneling-field-ionization on whisker-covered gold nanowires. IEEE Sensors Journal. 2008;**8**(2):161-169

[38] Abedini Sohi P, Kahrizi M. Low-voltage gas field ionization tunneling sensor using silicon nanostructures. IEEE Sensors Journal. 2018;**18**(15):6092-6096

[39] Edgcombe CJ, Valdrè U. Microscopy and computational modelling to elucidate the enhancement factor for field electron emitters. Journal of Microscopy. 2001;**203**(2):188-194

[40] Forbes RG, Edgcombe CJ, Valdrè U. Some comments on models for field enhancement. Ultramicroscopy. 2003;**95**:57-65

[41] Zou R, He G, Xu K, Liu Q, Zhang Z, Hu J. ZnO nanorods on reduced graphene sheets with excellent field emission, gas sensor and photocatalytic properties. Journal of Materials Chemistry A. 2013;**1**(29):8445-8452

[42] Charlier JC et al. Enhanced electron field emission in B-doped carbon nanotubes. Nano Letters. 2002;**2**(11):1191-1195

[43] Sohi PA, Kahrizi M. Formation mechanism of silicon nanowires using chemical/electrochemical process. IEEE Transactions on Nanotechnology. 2017;**16**(3):507-513

Chapter 8

Nanomaterials via Reconfiguration of Skeletal Matrix

Rajendra Sukhadeorao Dongre

Abstract

Nanotechnology is an ever-expanding field, which offers novel avenues due to advance, unique, and myriad applications in science and technology. Especially composites procured from reconfigurated matrixes appear as multiphase matter tendering augmented/new functionalities via chosen amalgamations. Hence, it is important to meticulously comprehend the interactive materials for the basic reconfiguration of their skeletal matrix to derive desired output to cater to the needs of S&T developments. Nanoscale material's systematic and rational designing gets fundamental as various material scale manipulations permit to recognize characters and functionalities that are not viable via conventional methods. Material's skeletal matrix reconfiguration is feasible through advanced biotechnology, physics, chemistry, and nanomaterial engineering mainly decisive to fabricate the particle, thing, and device at the atomic and molecular dimensions. These reconfigurations of material's matrix reduce its spatial dimension/captivity within crystallographic phase usually changing its physical, mechanical, thermal, optical, and electrical-electronic properties. Reconfigurated material matrixes restrain three nanoporous skeletons, namely: 3D/zero dimensional (e.g., particle, grain, shell, capsule, ring, and colloidal), 2D/one dimension (e.g., quasi crystal, nanorod, filament, tubes, and quantum wire), and 1D/two dimensional (e.g., disc, platelet, ultrathin film, super lattice, and quantum well). Today, rational designing of smart nanomaterials obtained via flexible matrix's skeletal reconfiguration focus on desired applications in the advancement of science and technology.

Keywords: nanomaterial, nanotechnology, reconfiguration, 1-/2-/3-dimensional, metal-organic framework/matrix, graphdiyne, graphyne

1. Introduction

Nanotechnology aids in designing challenging functionally fabricated materials via skeletal matrix reconfigurations that offer integrated solutions for medical, biological, and engineering model applications [1]. The hierarchical structure alterations open/allow nanomaterial creations with comparable size to those that occurred naturally. Accordingly, advanced engineering/technological designing permits assorted skeletal reconfigurations to yield resultant matrixes (also abet to mimic intricate natural skeletal) proposed certain advanced utilities viz. drug delivery, gene capture, tissue release, cell labeling, and scaffold improvements [2].

2. Nanomaterials: classifications, characterizations and fabrications

'Advanced nanotechnology' offers novel tools, opportunities and scope in significant applications for diagnostic and therapeutic purposes. Rationally designed smart materials are well established for drug delivery, diagnostics, treatment and prognosis of illness via reconfigurated/restructured devices and tools. Pharmaceutical nanotechnology is one of such fields embracing nanostructured products owing improved and requisite characteristics in its assorted sub-domains, viz. polymeric nanoparticle, magnetic/metallic nanoparticles, biosensors, biomarkers, liposome, carbon nanotube, quantum dot and dendrimers which are innovatory in medical/clinical usages [3]. Nano-pharmaceutically designed/reconfigured materials own vital applications in health risk-related issues like delivering therapeutic components at desired site in treatment of critical illness and crucial diseases. Nanoparticles are obtained via three techniques, viz. dispersion of preformed polymers, ionic gelation (hydrophilic polymeric coacervation) and monomeric polymerization. In order to control nanoparticle size/shape and composition industrially, certain methodology called supercritical fluid technology and particle replication in non-wetting templates [1–4] are also used to get reconfigurated nanomaterials. Dispersive polymeric technique is well used to reconfigure biodegradable nanoparticles via dispersion of biodegradable polymers like polyglycolide, polylactic acid, polycyanoacrylate and polylactide-co-glycolide [5].

Reconfigured matrix acts as effective carriers vulnerable in drug delivery, targeting cell release and specific tissue liberation in biological samples. Moreover, the size reduced targeted formulations; besides, suitable drug delivery pathway designing can be easily performed through these reconfigured matrixes.

Through intrinsic skeletal reconfigurations yielding varied nanostructures which displayed sole physicochemical/biological features. Rationally designed nanoparticles own potential utility for such purpose, imparting peculiar advantages like less toxicity, more release capacity, improved solubility, bioavailability and better drug formulations [3, 4]. Nanotechnologically achieved reconfigurations offer assorted nano-range matrixes owing to augmented performance. There are certain notable rewards of reconfigurations in material skeleton attenuated in nanosize matrixes such as reduce fed/inconsistency, amplified surface area, improved solubility, better bioavailability, more dissolution, fewer doses and rapid therapeutic action [4, 5].

Nanotechnology field in combination with advanced electronic, physics, and engineering sciences can offer superior applications in the domain of biophysics, molecular biology, medicine, immunology, cardiology, endocrinology, ophthalmology, oncology, and pulmonology, targeting brain and tumor besides gene/cell/tissue delivery [5]. Science of materials/devices can be restructured, and the resultant matrix reveals novel and substantially transformed nanoscales owing to unique physical, chemical and biological phenomena. Thus, nanotechnological reconfigurations in materials are manipulated at atomic, molecular and supramolecular level linking design, production and characteristic functionality anticipated for novel science and technological advancements [1–6]. Such reconfigurated matrixes are utilized in making molecular tools which preserves/improves our health through diagnosis, treatments and prevention of diseases, traumatic hurts and pain reliefs [1, 3]. The usage of assorted nanomaterials in the field of science and technology is mentioned in **Figure 1**.

Certain novel therapeutic and drug delivery systems are framed by the use of reconfigurated nanoparticles in medicines which also extend its utility for remedial and diagnostic research. Varied advance nanoparticulates like paramagnetic nanoparticle, quantum dot, nanoshell and nanosome get reconfigurated for cancer detection based on fluorescent material and contrast agent, targeting antibody besides molecular research which modernize and amend landscape of

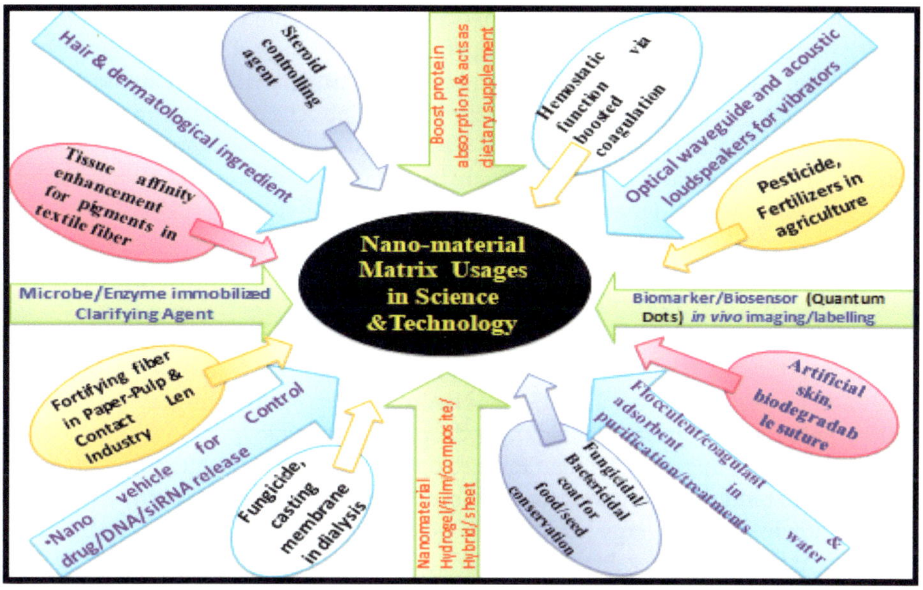

Figure 1.
Assorted usage of nanomaterials in the field of science and technology.

drug development pharmaceutics [1–7]. Assorted material reconfiguration yields products like liposomal and polymer-drug conjugates which carry active targets or perform controlled drug delivery as approved by the USA for clinical development [1–8]. Reconfigurated nanostructured materials with requisite features achieved via surface modifications and/or coatings with improved biocompatibility and bioavailability as employed in orthopedics, tissue engineering and dental for encapsulation, bone replacements, prostheses, implants and scaffolds appear better than conventional counterparts [9, 10]. Reconfiguration alters raw-material matrixes at atomic/molecular scale as viable for augmented particular shape, size and functional alteration in the form of quantum dot, biosensor, bio-detector, biomarker, dendrimer, nanocarbon, fullerene and nano-array [11–14].

2.1 Nanoparticle taxonomy

Nanoparticles are categorized in three classes: 1D nanoparticles, e.g. thin film 1–100 nm or monolayer/manufactured surfaces used in solar cell technology, chemical/biological sensors, information storage systems, magneto-optic and optical device and fibre-optics [1, 15]; 2D nanoparticles, e.g. carbon nanotubes; and 3D nanoparticles like dendrimer, quantum dot and carbon-60/fullerene [1, 7, 8, 15]. Nanoparticles are characterized through their size/shape, morphology and surface charge, by means of sophisticated microscopic systems, viz. atomic force microscopy, scanning electron microscopy, transmission electron microscopy, etc. [15]. Varied size distribution, average particle diameter and charges of nanoparticles are found to affect physical stability besides in vivo distribution. Electron microscopy gives information about its surface morphology, size and overall shape. Stability and re-dispersibility of the polymer dispersion and in vivo performance get affected by surface charge of reconfigured nanoparticles. Such nanoparticles characterized by assorted methodology are revealed. Nanoparticle size portrayal can be evaluated by particle size distribution and morphology, while electron microscopy ascertains both the morphology and size. Application of nanoparticles in drug release and drug targeting can be conveniently determined by various tools. Size of nanoparticle owns reflective consequence during the drug release applications. Small-size nanoparticles possess larger surface area and impart fast and significant drug release through drug carrying phenomenon as tiny particles get amassed during storage and dispersive transportation [16]. Thus the mutual compromise of upmost stability and pertaining small size is favored in reconfigurated nanosize materials [1–16]. Surface charge and intensity parameters decide electrostatic interaction of reconfigured nanoparticles with biological milieu or bioactive samples.

2.2 Reconfigured nano-colloidal

The mixture of microscopically disperse insoluble particles which are suspended all over another substance is called as colloids. Colloid mixture cannot settle down easily and takes time appreciably [17, 18]. Nature of interaction of dispersive phase and its medium differentiate colloids as hydrophilic/reversible sols and hydrophobic irreversible sols. The stable colloidal form remains suspended in solution at equilibrium, and it's hindered by aggregation and sedimentation as driven by innate tendency of reduction in surface energy [17, 18]. Low interfacial tension stabilizes such colloidal. Rheological functions are valuable in reconfiguration of suspensions in nano-colloidal forms as low viscosity and high shear rates aid deagglomerated mixing which governs suspension flow. Such colloidal state parameters are reconfigured for their elevated particle size, shape/flexibility and surface chemical-electrical properties. The colloidal dispersion obtained through reconfigured alterations that own suitable particle and medium interfaces offers well-defined large surface area as the best to be utilized for emulsion/oil-water separations. Such rationally reconfigurated interfaces tender facial capillarity, which is vulnerably essential for adsorption; for example, nanocarbon colloidal-based filters are used in purification of drinking water, beer/wine, decolorization of sugar, and gas masks [1].

Certain particle size reconfigurations have open influence on the bioavailability of active pharmaceutical ingredients and further effectively deliver intravenous lipid emulsion. Suspensions and colloidal dispersions own a range of sizes/shapes, and reduction of nanoparticle size can enhance surface area besides enlarging surface area/volume ratio. Shape is irrelevant, while surface area per mass of colloid scale nanoparticle owes a huge surface area resulting in superior adsorption achieved via interactive suspension rheology, coating and adhesion. Reconfigurated nano-colloidal permits faster dissolution of active pharmaceutical ingredients and corresponding bioavailability in hydrophobic membrane porous species. The active pharmaceutical ingredients establish incompetent treatment due to critical factors like innate low bioavailability, elevated cost, and toxicity; thus, they seek nanotechnology reformulated and reconfigured colloidal matrixes to cater for the advanced pharmaceutical applications. Reconfigurated interfaces of dispersive phase and their medium are negligible for colossal materials but dominant in colloidal being decisive in physicochemical alterations like surface chemistry and in toto system's reactivity [1, 3].

The colloidal size/shape for particle crystallization can be obtained through irregularity or asymmetric environment of nanoparticles determines its overall physicomechanical features as needed for intrinsic pharmaceutical applications. Colloidal sizes can exist as corpuscular like spherical/ellipsoid, laminar such as disc/plate and linear, viz. rod/needle. Globular proteins shape up with approximate spherically compact random coil configuration, while assorted active pharmaceutical ingredients occurs as rod/needle shape. Macromolecular bio-protein, polysaccharide and artificial polymer can be reconfigured as long thread/branch series colloidal owing to substantial mechanical potency and durability. The shape of colloidal can be reconfigured as large extended unidimensional strings up to rigid dense random coils which get manipulated by many factors like solution temperature, pH and nature of salt/electrolytes. Liquid-liquid interfaces, colloidal dispersions, metal sols like gold nanoparticles, lyophobic colloidal dispersions including polymeric-embodied APIs and ionic solid surface charge govern unequal dissolution due to crystal lattice anisotropy [3].

Till date numerous "nanotechnology-based" colloidals are reconfigured for environment, nano-medicine, healthcare and cosmetics. The use of ZnO/TiO_2 colloidal in creams attenuated UV-A/UV-B rays which induced sunburns and skin cancer; toothpastes own nano-hydroxyapatite to fill tooth cracks, and anti-aging products use nanocapsules. Nanosilver colloidals are reconfigured for injury healing/wound dressings. Nanosize zirconia-based hydroxyapatite colloidal acts as bioactive ceramics for orthopedic weight-bearing implants owing to advantageous sintering. Traditional pharmaceutical have delivered precise therapeutic to its accurate targets without side effects by means of these reconfigurated material-based nano-medicine. Colloidal stability is measured through its zeta potential gradient as a function of the outer Helmholtz plane and surface of shear which indirectly compute surface charges. Zeta potential in colloidal dispersions assesses storage stability as high potential either positive or negative certifies innate stability and avoids aggregation pivotal to evaluate surface hydrophobicity, encapsulation and surface coatings [1, 3, 17, 18].

2.3 Nano-colloidal analysis techniques

The establishment of colloidal surface hydrophobicity as achieved via sophisticated analytical techniques, viz. hydrophobic interaction chromatography, biphasic partitioning, and contact angle measurements. Rather X-ray photon correlation spectroscopy resolves surface hydrophobicity, which also aids in recognizing certain surface proactive functionalities [1, 3]. UV HPLC, ultracentrifugation, ultrafiltration, gel filtration, centrifugal ultrafiltration analyses and the extent of drug release parameter require its delivery being a vehicle like role of reconfigured nanomaterials [3, 17, 18].

2.4 Nano-dendrimers

Dendrimer is a unique polymer owing to a quite manageable size/shape with compartmental zones consisting of hyper-branched and tree-like reconfigured skeletons. The convergent or divergent step growth polymerization approach is used in fabrication of dendrimer skeleton from its monomeric units. Regular-branched polymeric nanostructures of dendrimers' own size depend on its controllable branching quantification [3, 19]. The nanostructure dendrites right from core shape alterations to spherical arrangements can also impart cavities during synthesis polymerization and are achieved through numerous reconfigurated alterations. Reconfigured dendrimers are effectively utilized in drug transport via its free ends that are involved in conjugation or attachment. End groups of dendrimers are tailored through interconnecting networks as per needful conditions which can transport loaded moiety/drug at desirable site to impart novel biomedical applications [1, 19]. Dendrimers have fine nanostructures and are also capable of surface functionalization; monodispersions with immense stability make it smart carrier for drugs. Drug moiety undergoes complexation and encapsulation in such dendrimers at its basic core, branches and surface skeletal units. Branch/end groups tailoring or grafting into biocompatible and high biopermeable species networking aid sustainable delivery of vaccine, cell, drug, gene and metal. Reconfigured dendrimers' own innate hollow arrangements with space/voids to include drug/bioactive samples through physic-chemical interactions help its control delivery due to beneficial features like modulated target-specific delivery, feasibly defined molecular weight, good trap capacity, flexible surface functionalization and lowest polydispersity index. Poly(ethylene glycol), chitin, melamine, polyglutamic acid, poly-propylene imine, polyamidoamine and polyethyleneimine biodegradable skeletons are easily reconfigurated into dendrimers via the above-mentioned synthetic methods. Dendrimer-based complexes act as nano-device's own potential utility in cancer chemotherapy as targeted drug therapy like tecto-dendrimers owing to every dendrimer unit exhibits assorted role, e.g. targeting, disease diagnosis, drug carrier and imaging [14]. Reconfigurated dendrimers used in therapy avoid stimulated immune side effects. Drug/therapeutic dendrimers conjugate target cells and indicate useful advantageous features, viz. boron EGF-carrying PAMAM dendrimers, intra-tumoural injection and CED-doxorubicin-2,2 bis(hydroxymethyl)propanoic acid-derived dendrimers, showing in vitro/in vivo less toxicity in colon carcinoma cell treatment in rats [1, 3, 17, 18]. Cationic dendrimers showed more cytotoxicity, cell membrane instability and cell lysis than anionic dendrimer, PAMAM. Assorted nano-dendrimers' yield via block polymerization and chemical cross-linking is shown in **Figure 2**.

2.5 Nanocrystal and nanosuspension

Aggregated structures are formed through the combination of various particles in crystalline form coated with surfactant combinations which impart static and electrostatic surface stabilization. Such nanocrystal and nanosuspension aggregated materials lessen bioavailability and absorption issues due to its resolved formulation. Nanocrystal size permits safe and effective passage via the capillaries. Reconfigurated solid lipid nanoparticles owe discrete reward of unique carrier systems over liposomes and polymeric nanoparticles due to solid lipid matrix owing to 1 μm diameter where drug gets easily incorporated. Such nanoparticle/nanosystems for clinical usage are obtained by high-pressure homogenization that uses varied surfactants to avoid aggregation and imparts dispersion stability. Reconfigurated cationic solid lipid colloidal nanoparticles are prepared and used for liposomal transfection agents besides gene transferring dominant for in vitro [1, 3, 15–18].

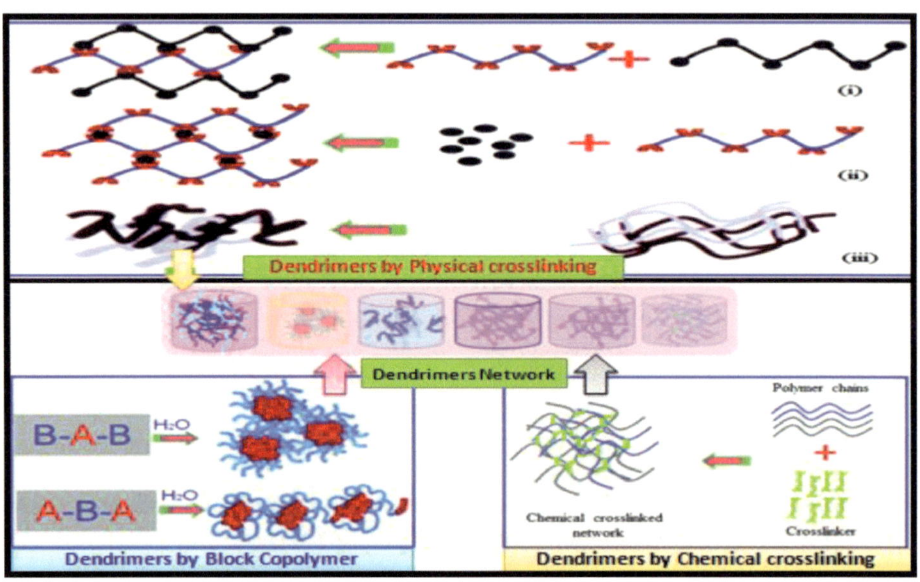

Figure 2.
Nano-dendrimer synthetic route: block polymerization and chemical cross-linking.

2.6 Silicon-based nanomaterials

Silicon-based nanostructures are reconfigured by means of assorted techniques including etching, photolithography and deposition. Materials like porous silicon, silica and silicon dioxide are facile to reconfigure as the calcified nanopores, nanoparticles and nanoneedles. Assorted porous hollow silica nanoparticle suspensions owing sacrificial nanoscale templates mostly involved the usage of sodium silicate as a precursor. Such reconfigurated silicon-derived materials can offer effective delivery of drugs like porous silicon-embedded platinum which is applied for many usages like antitumour drug release and carrier for antibody, antibiotic, enzyme and DNA. These materials also act as good semiconductors and thus preferred in micro-electromechanical usage [1–3].

2.7 Nano-micelles

Polymeric micelle assemblies are reconfigured through amphiphilic block copolymers which impart nanoscopic supramolecular core shells as aggregates in solutions. In such micelles, the reconfigured components get ordered as spheroidal owing to hydrophobic core shells that are water secured due to hydrophilicity. Many components owing to hydrophobic and hydrophilic portions are reconfigured like amphiphilic AB, or ABA kind of block copolymeric units yield polymeric micelles as nanoscopic supramolecular core shells as aggregates in solutions [1–3]. Nanoscience technology has reconfigurated assorted supramolecular architectures as a result of self-assemblage of amphiphilic block polymers through hydrophobic/hydrophilic effects, electrostatic interactions, hydrogen bonding and metal complexation which proffer sharp structures and precise functionalities for their usages in biomedical purpose. Such polymeric micelles fascinated special attention in drug/gene release by virtue of outstanding biocompatibility, less toxicity, more blood circulation and enhanced solubility in its innate micellar core shells. Polymeric micelles act as 'smart drug carriers' due to binding of specific ligand onto proactive surfaces which aids targeted/formulated stimuli sensitivity found devoid in their counterparts. Self-assemblage in block/graft copolymers occurs in selective solvents spontaneously reconfigured as supramolecular assemblies owing to cylindric and vesicular core shells of 10–100 nm shape/size nanomaterials called polymeric micelles. Spherical reconfigured core-shell size/morphology gets crucially varied with chemical structures and compositions of constituent block copolymers besides exhibited elevated thermodynamic/kinetic stabilities over surfactant micelles. Polymeric micelle yields through advanced reconfigurations own an external hydrophilic nanoshell that assists in prolonged residency in blood, besides gathering in tumor-specific zone due to more permeation, and thus acts as a novel vehicle for sparingly soluble (hydrophobic) drugs. Polymeric micelles have shown adaptability to various ligands which aids active targeted delivery by virtue of better solubility and improved pharmacokinetics besides no adverse effects. Great interest is generated by these micelles in technological advancements due to remarkably stable and bendable physicochemical features exhibited with various stimuli [1–3, 19].

2.8 Biopolymeric nanoparticles

Nano-array of polymer-based nano-conjugates obtained from different natural sources own modern functions besides specific and targeted drug delivery. The natural polymeric reconfigured nanoparticles are biocompatible, less toxic across many bio-membranes against various pH and non-immunogenic and appear to be extra stable to volatile pharmaceutical agents besides offering low-cost fabrication at large-scale/multitude methodology. Biopolymeric reconfigured nanomaterials are biodegradable as anticipated for tissue engineering scaffolds and drug/gene release/ carrier besides novel vaccination approach [3, 19]. Certain natural polymers like chitosan, gelatin and sodium alginate owing to nontoxic profile get easily reconfigurated in situ with man-made polyesters, viz. polycaprolactone, polycyanoacrylate, poly (lactide-co-glycolide), polylactide and polylactic acid, to yield resultant matrixes advocated under advance nano-biotechnology. These bio-polymer-derived nanoparticle offers to develop utility over usual oral/intravenous mode of drug delivery with more competence and efficiency. Certain polymeric nanostructures possess homogeneously dispersed template which can be reconfigurated as vesicular systems like nanocapsules and matrix systems like nanospheres. Reconfigured nanocapsules owe innate cavities that can detain assorted biomoiety including drugs, genes and cells enclosed by polymeric membrane, while nanospheres biomoiety get diffused all over its polymer matrix. Polymeric nanoparticles adopt as globular vesicular nanocapsules wherein polymeric membrane can dissolve, entrap and attach/encapsulate foreign moiety throughout its reconfigured core-matrix/skeleton. Polymeric background is chosen which owns ability of modifications so that resultant nanoparticles can act as ideal carrier/vehicle for delivery of assorted species, viz. drugs, vaccine, contraceptive and antibiotic. Reconfigurated polymeric nanoparticles act as an attractive module for intracellular and site-specific delivery besides engaging in fabricating smart scaffolds/templates in advance tissue engineering [20–23].

2.9 Nano-graphynes

Carbon owing to versatile Dirac cones and hexagonal networking is called as graphyne which exists as α, β and γ forms. Theorized allotrope of carbon called graphyne is found to own carbon-carbon triple bonds pertaining versatile Dirac cones, which aid to perform an astonishing role in reconfigured atomic/electronic structural materials [1, 24]. Such carbon triple bonding adapts hopping template elements with undo signs which yields Dirac cone with perverse chirality in reconfigurated α, β and γ graphynes which impart momentum shift of energy gaps besides offering chemisorption of adatoms via sublattice symmetry loss. Unique characteristics of such 2D carbon nanomaterials can be reconfigured as graphyne and stacking graphdiyne found to stimulate innovative and fascinating utilities in advance electronics. Atomically specific 2D graphdiyne and graphyne matrix reconfiguration is an awaited challenge for material scientists. Technique of on-surface synthesis in ultra-high vacuum yields graphyne that can be further feedstock for making comprehensive graphynes in particular atomically precise graphdiyne nanowires. In the past decades, low-dimension carbon materials, viz. fullerenes, carbon nanotubes and graphene, have ever-fascinated scientific and technological focus. Amid two-dimensional carbon, allotropes called graphene are deliberately pioneered via morphological reconfiguration as porous nano-strips or nanoribbons which own innate bandgap as competitive/superior as graphene. Graphynes are also tentative artificial carbon allotropes owing to intervallic acetylene bondings with blend sp^n hybridization where $1 < n < 2$. Thus based on sp./sp^2 hybridized carbon, such graphynes are categorized as α, β and γ and named as graphyne, graphdiyne and n-graphyne where $n > 2$, as per acetylene units [1, 24].

Reconfiguration of single layer, i.e. 2D carbon allotrope graphyne, is still an exigent task, since hardly any natural crystalline substances contain stacked graphynes as inclusive in auxiliary nanostructures like nanotubes, nanoribbons, quantum dots and junctions. The graphyne's innate C-C triple bonding is facile for advanced reconfigurations via attachment of hydrogen/halogen without disturbing innate two-dimensional hexagonal planarity; such reconfiguration tunes energy gap factor at the Dirac point [3, 24]. These graphynes' synthesis is achieved through realistic and choosy functional polymerization owing to diligence precision up to atomic scale, e.g. dehydrobenzoannulenes onto catalytic copper foil via acetylenic cross-coupling resoluted core-graphyne subunits. Dehydrobenzoannulene can also be reconfigurated in assorted morphologies including single-layer graphdiyne, tubes, wires and walls which thrust significant utilities in catalysis and energy field.

2.10 Reconfigured graphdiyne nanowire

Nanoscale graphdiyne-derived templates are developed via Glaser reaction/ acetylenic homocoupling to yield targeted π-conjugated 2D nanomaterials which further can be reconfigured as sp-hybridized nanostructure matrixes [3, 24]. Certain terminal alkynes like 1,3,5-triethynyl-benzene act as convergent tritopic precursors in reconfiguring graphdiyne-based porous matrix via mild thermal annealing. Discriminating butadiyne inspires elementally incarcerated graphdiyne reconfiguration in the form of nanoribbons and *m-n* nanowires, where *m* is phenyl rings and *n* is alkynes through the recurring backbone (**Figure 3**). Once side functionality gets established in the graphdiyne moiety, it improves its quality in extended polymerize nanowires, which are best utilized in augmentation of molecular electronic parameters. Graphdiyne nanowires in vacuum own an energy gap of $\approx$1.6 eV; further statistical twisting of phenylene indicated fine changes in electronics due to cosine highest valence band and lowest conduction band viable for nonlinear electronic transportations like Bloch oscillations appropriate in high-frequency tools. Graphdiyne nanowires proffer notable automatic strength and elasticity if acetylene bondings get well conserved and offer constant chemical characteristics. Superior grade graphdiyne nanowires are prepared using butadiyne precursor through assorted tactics, viz. thermal processing, substrate selectivity, molecular designing, surface templating and metal-organic bonding creations. Raw feedstock selection is crucial in reconfiguration of π conjugated 2D nanomaterials like graphyne and graphdiyne derivatives. An atom that lies on a surface of crystal acts as the reverse of a surface vacancy and is called as adatom, and it can be cited/reconfigurated onto the top layer of metal surfaces, which impart proactive shell seeking the best adsorption molded molecular deposition and distinguished catalytic properties. On-surface acetylenic glacial coupling using silver metal is suitable to get acetylenic linkages in resultant graphyne and graphdiyne derivatives. Copper and gold both are primal metal for alkyne homocoupling with ditopic 1,4-diethynylbenzene as its over-reactivity gives extra reactions. Gold displayed surprising cyclotrimerization depending on the symmetry of precursors like if three terminal alkynes get mutually coupled to form benzene.

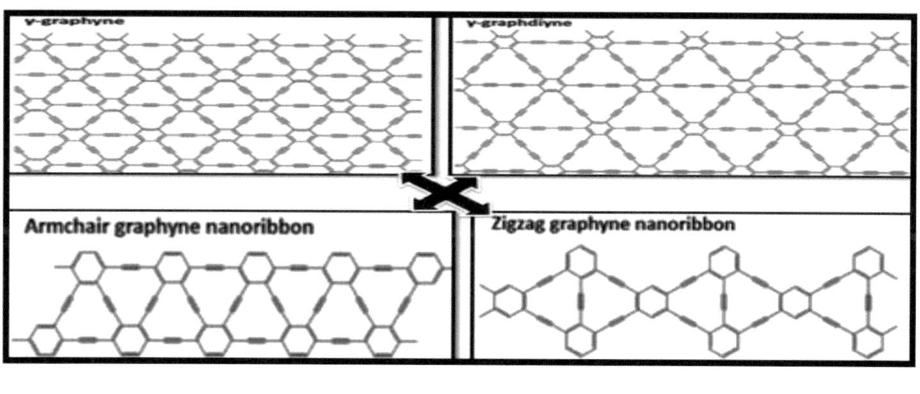

Figure 3.
Schematic representation of graphyne and graphdiyne matrixes.

Lower-mobility species gets easily detained onto metal which resulted in side reactions in concurrence to alkynes' cyclotrimerization. Ortho functional groups restrain tangential terminal alkyne contacts which results in butadiyne linkages via weak supramolecular interactions amid nanowires that can influence neighboring molecular alignments. Anisotropic motif in asymmetric 1,2,4-cyclotrimerizations over symmetric 1,3,5-cyclotrimerization onto gold surfaces is preferred to get H-shaped oligomer and intrigue alkyne-gold interactions. Graphyne and graphdiyne derivatives are advantageous than mere graphene for innate electronic features [1, 3, 23]. On-surface acetylenic couplings can expand graphyne and graphdiyne networkings which correspondingly distinguish in intrinsic physic-chemical characters. Linear expansion of graphyne and graphdiyne into 2D graphyne and graphdiyne derivatives is still exorbitant, so strategic halide usages avoid influence of hydrogen abstraction forming hexagonal planes on gold bridging mutual acetylenes. Hexaethynylbenzene is mostly used for getting reconfigured single-layer stacking of graphyne/graphdiyne via acetylenic couplings. Advance on-surface synthetic protocols are developed for graphyne-/graphdiyne-based atomic precise nanowires, quasi-1D nanoribbons and 2D networkings. Porous structural reconfigurations of 2D materials are noteworthy for innate electronic and mechanical usages [1, 3].

2.11 Reconfigured metal-organic frameworks

Recently carbon based from metal-organic frameworks is more demanding due to innate advantageous electron conductivity and extra porosity sought in diverse fields. Thus morphology of carbon materials gets improved through altering its chemical/physical characteristics via optimized compositing with metal-organic materials. Such carbon-reconfigured metal-organic frameworks appear nontoxic and offer brilliant electrical conductivity in contrast to other energy storage materials. This makes carbon-based metal-organic reconfigurated frameworks superior to most of the energy storage materials offering promising functions in demand for energy storage/conversion and addressing confronts in lithium/lithium-sulfur/sodium battery, metal-oxide/sulphide-carbon-based supercapacitors and electrocatalytic oxygen/hydrogen reduction/evolution reactions besides water-wastewater treatment techniques. Ultrasensitive biosensors are reconfigured through carbonaceous skeleton via N/S doping to get electrodes for in vitro monitoring of uric acid and ascorbic acid released from living cells. Direct physical/chemical carbonization of organic templates with assorted species like zeolites, meso-silica via solvothermal or hydrothermal techniques yields carbon-based metal-organic frameworks [25]. Carbon reconfigured through metal-organic porous coordination polymers resulted in crystalline porosity due to episodic metal ion/metal clusters with organic ligand networkings. Metal-organic reconfigured carbon matrix/framework sustains assorted reward like extra porosity with tuneable sizes and very high up to 10,000 $m^2\ g^{-1}$ surface area as anticipated for adsorption, energy/gas storage/conversion, oil-water separation, catalysis, markers/sensors and solid-phase extraction. Rationally hierarchical porous nitrogen-doped carbon frameworks are developed through zinc and nitrogen templates which are used for elevated storage and adsorption capacity for CO_2 gas. Iron, zirconium and lanthanum metal-doped-NH_2 reconfigured frameworks are obtained via solvothermal process followed by pyrolysis, yielding nanocarbon matrixes which are used for biosensor activity [1, 3, 25].

2.12 Reconfigured carbon nanocages

Reconfigured nanocarbon cages can act as electrode; for example, hollow nanocobalt sulfide intervening graphitic nanocage offers superior lithium storage capacity along with stable performance in advanced batteries. Porous ZnO-carbon nanocage is reconfigured through pyrolysis of hollow MOF-5 owing to high specific surface areas besides hollow morphology [1, 3]. Porous Co-Zn-NH_2-doped carbon polyhedral nanocage effectively acts as anode in lithium-ion batteries. Transition metal oxides of M_xO_y types derived from Mn, Fe, Ni, Co and Sn offer superior capacities of 1000 mA h g^{-1} than that of graphite templates and thus gain weightage as anodes in sodium-ion batteries. However, transition metal oxides of M_xO_y types can be reconfigured in carbon materials so as to proffer elevated surface area and improve sodium storage simultaneously. Hollow NiO/Ni nanocrystal reconfigured onto graphene shell imparts good storage capacity and cycle stability [1, 3–25].

Carbon matrix gets reconfigured with nitrogen doping in Co_3O_4-based metal-organic hybrid/framework imparting unique features like high electronic conductivity, superb definite capacity and superior cycled constancy. Many bimetallic Ni-Co-organic frameworks owing to hierarchical hollow crossbreed occur via generic template-free strategy to fabricate anode electrode for sodium-ion battery. Hollow nano-skeleton of such organic framework/hybrids reconfigured entirely novel electrodes' utility owing to constant reversible capacity after long-term 200 cycles. Titanium-derived metal-organic hybrid crystal yield via carbon-coated rutile materials is used for anode making in sodium-ion battery up to 2000 cycles. Graphene-titanium oxide reconfigurated metal-organic hybrids/composites can act as anode found to exhibit huge porosity and great retention capacity up to 5000 cycles for sodium-ion battery usage [11, 24].

Transition metal sulphide-based carbon hybrid like Ni_3S_2/Co_9S_8/N–C-gifted hollow-spheric skeleton obtained via carbonization-sulfurization of binary Ni–Co metal-organic framework is used in fabrication of electrode for sodium-ion battery. Ultra-fine hollow porosity is achieved in nanometal sulfide blended ultrathin N-doped organic carbon hybrids, which delivered brilliant electrochemical function with competent capacity up to 100 cycles. Bimetallic zinc antimony sulphide which blends organic carbon core double shell polyhedron frameworks exhibits unique electrochemical functions like consistent cycling stability and elevated coulombic efficiency besides precise capacity. Ultrathin/nano-molybdenum sulphide coated onto flexible N–C/carbon cloth nano-array hybrid/sheet owns good electrochemical performance as an anode up to 1000 cycles for sodium-ion battery. Such admirable electrochemical functions are accredited to unique two-dimensional features viable for curtail ionic diffusion which overall extend Na^+ insertion. The N-doping in organic carbon yields porosity in resultant nanowall which imparts advance conductivity and sustainable integrity of such bimetallic organic frameworks [1–3, 23].

Organic/carbonaceous nitrogen-doped metal porous frameworks get reconfigured for making sodium-ion battery electrodes' residing unique features like fair capability, cycling stability, high electrical conductivity and elevated ion storage ability. Inorganic sulfur/phosphorus reconfigured with metal-organic template aids to fabricate S/P-doped meso-carbon anodes which are practically reported to prolong cycle constancy, elevated energy density and wonderful rate capacity for sodium-ion battery. Amorphous red phosphorus reconfigured into micro-porous carbon-nitrogen matrix shows high Na^+ storage performance and reversible capacity up to 1000 cycles for sodium-ion battery. Certain nano-sheets get reconfigured through 3D reduced graphene oxide-anchored phosphorus-nickel foam owing to cobalt core shell of phosphorus-carbon polyhedron which overall improved cycling stability and benefited damage relaxation during charging-discharging in resultant battery electrode performance. These reconfigured materials hold sole metal-organic framework as electrode which showed astonishing and superior electrochemical outputs like high power density, reversible capacity, brilliant stability, huge cycling stability, elevated rate capacity and galvanostatic charge-discharge contour. Assorted homogeneous-dispersed hierarchical 1-/2-/3-D porous micro-/nano-layer cathode matrixes are reconfigured with nitrogen-doped organic frameworks through melting-diffusion and infiltration techniques. Extensive metal-organic frameworks are obtained via adjusting morphology, reactive conditions, and control carbonized reconfigurations in feedstock materials and are found to proffer characteristic features like greatest recycling, prolonged cycle electrical capacity, and top capacitance as desired in advance battery electrode fabrication [3, 25]. Doping of nitrogen/sulfur further raised charge redistribution, electron delocalization and reversible capacity in resultant batteries that are superior to other counterparts. Metallic insertion in carbon frameworks yields copious and consistent porosity in ultrafine nano-skeletons as beneficially immobilized N/S in organic templates. Similarly, sodium-ion battery electrodes are reconfigured via phosphorus/nitrogen added to carbon scaffolds so as to deliver superior electrochemical functions and impart high theoretical sodium-storage capacity due to Na-P synergistic effect devoid in its contemporary [1–3].

Nanocarbon-reconfigurated metal-organic frameworks are superior to fabricate electrode with improved electrochemical performance in supercapacitors, e.g. zinc-organic scaffolds. $Li_4Ti_5O_{12}$-derived hybrid super-capacitor yields Li-type anode and double-layered cathode electrochemical capacitance with duly great energy density and prolongs capacity up to 10,000 galvanostatic cycles. Reconfigurated hierarchical carbon-coated tungstic anhydride-Li-HSC porous anode and N–C hollow polyhedron-based cathodes exhibited soaring power density and high-retained capacity up to 3000 operating cycles [20–26].

Splendid 'brick-and-mortar'-type squash-in nanoporous matrixes can be reconfigured owing to dope metal as 'mortar' and abide organic framework as 'brick' to be used for electrode in next-generation energy/power storage battery. Remarkably reconfigured novel nanocarbon metal-oxide sheets/core-shell composite gets casted as asymmetric supercapacitor anodes owing to longer cycle stability, high energy density and huge energy density in aqueous electrolytes with upmost 10,000 cycle capacitance retention superior to other counterparts. Nanocarbons reconfigured onto bimetals yield hybrid via control carbonization, and coexisting N-doping architecture imparts huge surface area and better capacitance (than nanocarbon) to be used to fabricate flexible asymmetric advanced supercapacitors [3]. Unique one-dimensional hollow structure of N-dope-organic framework-reconfigurated bimetals imparted good power storage capacity and electrochemical performance up to 10,000. Metal oxides derived from supercapacitors conveyed superior power density and electrochemical stability than carbonaceous polymeric matrix. Flexible hybrid supercapacitors developed from N–C-doped niobium-oxide quantum dots performed as superior electrochemical electrodes owing to maximum energy/power density and cyclic stability even up to 5000 cycles [27]. Porous manganese tetroxide reconfigured N-doped graphene by means of polystyrene template yields ordered porous composite owing to brilliant electrochemical capacitance and cyclic stability up to 2000 cycles in aqueous electrolyte solutions. Ruthenium oxide reconfigured nanocarbon scaffolds have shown outstanding electrochemical performance as supercapacitors owing to capacitance retention in lithium-ion battery. Assorted hierarchical nanoporous organic carbon-based skeletons are developed for its innate higher capacitance and cyclic stability even up to 140,000 cycles than bulk amalgamated electrodes. Thus, reconfigurations of material indeed have to break through many challenging discoveries of advanced anode/cathode materials in the development of high-performance batteries owing to good volumetric/gravimetric energy density and its allied futuristic functioning for electrification of vehicles besides grid power storage.

Noteworthy R&D is performed to discover high volumetric/gravimetric energy density electrodes through reconfiguration of material's matrix for getting innate electro-voltaic functions. Certain layered mixed metal oxides, viz. $LiMn_2$, olivine $LiFe\text{-}PO_4$, $LiCoO_2$ and $Li\text{-}Ni/Co/MnO_2$, are reconfigured to fabricate electrodes for lithium-ion cell. Strategies are being developed for optional high-energy cathodes with preserved substantial stability, rate capability and its cycle life. High-energy cathodes made from $Li2MnO_3$ conveyed electrochemical steadiness with improved specific capacity and corresponding volumetric/gravimetric energy density. Li-ion batteries offer prospective high volumetric/gravimetric power density of its cells achieved through reconfiguration of novel electrochemicals [26, 27]. Sulfur- and oxygen-based bimetallic organic framework-derived cathodes are intensely reconfigured for innate superior theoretical capacity over usual metal-oxide electrodes. The designing and development of progressive high-performance material-based electrodes derived via assorted reconfigured materials are still a difficult task/ challengeable due to their innate size constraint. Lithium-ion battery needs prospective high-power electrodes as derived via various reconfigured matrixes including Si-alloy, metal-oxide frameworks and graphene carbons for electro-vehicles and grid power storage. Certain reconfigurated material matrixes convey exceptional features like facile mount skeleton, high specific surface area and storage capacity besides hierarchical porosity contributed in innate onset potential electrochemical performances.

2.13 Environmental applications of MOF

Assorted metal-organic frameworks (MOF) are reconfigured to impart special valid features like huge specific surface area, adaptable porosity, and constitutional uniformity and to unlock metal sites although certain facile physic-chemical variations are anticipated in advanced S&T applications [3, 19] can be even utilized for better adsorption of dyes, gases and environmental pollutants as mentioned in **Table 1**.

Diverse materials are reconfigured to be used as adsorbents for mitigation of water pollution; for example, 3D twofold zinc-doped carbon porous scaffolds owing to elevated surface area conveyed five-fold higher sorption capacity for dyes and drugs like ibuprofen/diclofenac contaminated water over commercial-activated carbon [3]. Magnetic carbon sponges are reconfigurated in zeolitic imidazolate framework-67 to carry out excellent separations of buoyant oil from water and oil from emulsions and executed excellent catalysis in H_2 gas generation [25, 27]. Extremely dispersive nano-chromium oxide can be reconfigured in mesoporous carbon nitrides to yield MIL-100(Cr) templates owing to greater specific surface area which aids huge CO_2 adsorption capacity quite higher than its counterparts [1–3, 28].

Specially reconfigured nickel oxide/poly-carbon nitride interlinked with tree-like chains/branches owing to unique features like nano-flower/leafy planes, huge surface area and hirsute dendrite core shells; superior porosity can impart superior and control/choosy arsenate anionic diffusion besides efficient As^{+3} to As^{+5} oxidative conversion in contaminated water. Porous nano-spherical scaffolds holding iron-EDTA ligands owe exclusive chelating sites that afford huge anionic adsorption capacity of 307 and 407 mg g^{-1} for As (V) and Cr (VI), respectively. Metal-organic frameworks are used to reconfigure many fluorescent sensors/ markers like super-porous chemosensors owing to zirconium-based hydrophobic fluorescent probes developed to check ultratrace (0.1–2000 ppb level) Zn^{2+} ions from water. Ratiometric fluorescent sensor containing UiO-66-zirconium matrix is used for selective Zn^{2+} detection from water. Hydrophobic fluorescent probes reconfigured with rhodamine ethylene-diamine salicylaldehyde are developed for sensitive Bi^{3+} adsorption from water. All such reconfigurated metal-oxide carbon

Metal in organic frameworks	Example	Brunauer-Emmett-Teller (BET) surface area $[m^2\ g^{-1}]$	Pore volume $[cm^3\ g^{-1}]$/d	Absorption utility
Iron	Fe_3O_4-carbon	439	–/0.80 nm	—
Zinc	Hierarchical carbon-C1000	524	–/0.80 nm	59.20 $cm^3\ g^{-1}$ for CO_2
Zeolitic imidazolate framework-67	Magnetic carbon sponges	62	0.20/–	58,000 mg g^{-1} for silicone oil
ZIF-8	Iron-doped carbon	606	–/5 nm	Reflection loss of 29.50 dB
Ni-based MOF	Nickel-doped carbon	110	–/1.80 mm	Reflection loss of 51.80 dB
ZIF-8	PCDM-800 nanoporous carbons	885	–/–	–
ZIF-8	PCDM-1000/1200	1855	–/–	320 mg g^{-1} for IBP
Nickel/zinc	Ni-phthalocyanine -carbon nanotubes	999	0.90/–	395 mg g^{-1}, 898 mg g^{-1} and 271 mg g^{-1} Rhodamine-B, malachite green and for methyl orange, respectively
Materials of Institut Lavoisier MIL-100 (chromium)	Cr-MCN nano-chromium oxide-mesoporous carbon nitrides	1294	0.40/–	16.8 mmol g^{-1} for CO_2 and 22.5 mmol g^{-1} for hydrogen gas

Table 1.
Metal-organic framework-reconfigured carbon materials for adsorptions.

frameworks own wide utility in water treatment techniques like selective and sensitive adsorption/detection of pollutants from water/wastewaters. These reconfigurated matrixes, scaffolds and templates own magnificent adsorption profile due to remarkable features, viz. huge surface area H-donor/π–π bonding capacity and great hydrophobic environments [1–3].

2.14 Biological applications of MOF

The outstanding properties of reconfigured MOF matrixes are used for sensing or absorption of assorted biological species like genes, cells, tissues, drugs and selective single-stranded DNA and quenching label fluorescent dyes. Magnetic nanocarbon porous scaffolds are reconfigured via thermolytic iron oxides MIL-88A matrix to proclaim sensing platform for double-stranded DNA with target DNA and aids in the release of single-stranded DNA probes (1×10^{-9} m detection range) onto adsorptive surface in DNA hybridize sensor electrochemical impedance detection [3, 4]. Iridium-MOF-8 scaffolds act as hybrid electrochemical sensing/detection of anesthetic lidocaine (0.20×10^{-12}–8×10^{-9} M). N-link glycan-metal organic-based frameworks obtained through pyrolysis can remove zinc ions from soils and complex biological samples.

Challenging areas in MOF.

These reconfigured scaffolds, templates and matrixes yield advanced materials which fulfill existing needs like high-performance energy-power storage, adsorption and chemosensor/marker utilities in today's modern science and technology [1]. Reconfigurated materials offer superior benefits over traditional porous materials due to huge specific surface area, tuneable porosity and facile skeletal functions. Such innovative materials emerge through reconfigurated synthetic methods and material structures, and analogous outputs really trust reciprocated improvement in various disciplines including batteries, supercapacitors, electro-catalysts and water treatment techniques. Yet, few perspective challenges and key problems as presented below are to be solved:

- Synthetic protocols involved in reconfiguration of metal-organic frameworks need to be explored for applicability at high-temperature calcinations and to aid ordered annihilation and impart huge porosity besides carrying out effective polymerization.

- Metal-organic framework-reconfigurated matrixes still rely on material synthesis and its characterization, but R&D mechanistic performance is yet to be done. Methodical perceptions of interfacial interactions amid metal-organic frameworks and its doped components might offer superior optimized synthesis protocol.

- Strategic reconfiguration augmented MOF-doped polymeric adhesives, and control functionalized carriers are to be analyzed for its cost and operating strategy so as to boost electrochemical performance to meet industrial applicability.

- Meritorious MOFs need to be reconfigured as per adaptable structures, composition and morphology for its excellent performances.

- Experimental and theoretical computational simulations can give momentum for innovative reconfiguration as applicably predicted to fabricate advanced scaffolds, templates and matrixes via MOF.

- Rational design is to be done to strictly control structural, functional and performance-based aspects of metal-organic reconfigured materials.

3. Summary

Innovative and truly applied nanomaterials can be reconfigured through corresponding skeletal/matrix alterations which are successfully accounted in the past decades. Matrix reconfigurations in nanomaterial skeleton found to vary with environmental conditions, diverse elements, allied constituents and corresponding organic supports communicate intricate besides mixed chemical characteristics. Each of such modules owns crucial status, plays a vital role in its reconfigurations and harmonizes mutual deficiency besides proffered synergistic recitals. Usually, reconfigurated nanomaterials acquired expedient features like hierarchical porosity, superior mechanical stability and elevated electrical/thermal conductivity. Due to their promising features, reconfigurated nanomaterials are intensively used for many purposes including batteries, supercapacitors, electro-catalysis and water treatment techniques. Abundant notable advancements are attenuated performance utility which persuades novel breakthroughs in reconfiguration of futuristic nanomaterials.

Acknowledgements

The author is thankful to the Head, PGTD of Chemistry, R.T.M. Nagpur University, Nagpur, for laboratory facilities, and to the Vice Chancellor, Nagpur University, Nagpur, for the sanction of research project work under the University Research Project Scheme, awarded vide No. Dev/RTMNURP/AH/1672 (9) dated 24 September 2016.

Author details

Rajendra Sukhadeorao Dongre
Department of Chemistry, RTM Nagpur University Campus, Nagpur, Maharashtra, India

*Address all correspondence to: dongre1974@gmail.com

References

[1] Dongre RS. Biological Activities & Application of Marine Polysaccharides. Vol. 1. Croatia: In-Tech Open Publications; 2017. pp. 181-206. ISBN: 978-953-51-2860-1. DOI: 10.5772/65786

[2] Bracco G, Holst B. Surface science techniques. In: Lytken O, editor. Adsorption Calorimetry on Well-Defined Surfaces. Berlin, Heidelberg: Springer-Verlag; 2013. pp. 35-55. Chapter 2. ISBN 978-3-642-34243-1. DOI: 10.1007/978-3-642-34243-1

[3] Dongre RS. Chitosan-Derived Synthetic Ion Exchangers: Characteristics and Applications. Vol. 1. Croatia: In-Tech Open Publications; 2018. pp. 21-42. ISBN: 978-1-78984-248-7. DOI: 10.5772/intechopen.78964

[4] Jain T, Kumar S, Dutta PK. Chitosan in the light of nano-biotechnology: A mini review. Journal of Biomedical Technology and Research. 2015;**1**(1):101-107

[5] Bănică F-G. Chemical Sensors and Biosensors: Fundamentals and Applications. Chichester, UK: John Wiley Sons; 2012. p. 576. ISBN: 9781118354230

[6] Humplik T, Lee J, O'Hern SC, Fellman BA, Laoui T, Wang EN. Nano-structured materials for water desalination. Nanotechnology. 2011;**22**(29):1-19

[7] Burkatovskaya M, Tegos GP, Swietlik E, Demidova TN, P Castano A, et al. Use of chitosan bandage to prevent fatal infections developing from highly contaminated wounds in mice. Biomaterials. 2006;**27**:4157-4164

[8] Kurita K. Chitin & chitosan: Functional biopolymers from marine crustaceans. Marine Biotechnology. 2006;**8**(3):203-226

[9] Mourya VK, Inamdar NN. Chitosan-modifications and applications: Opportunities galore. Reactive and Functional Polymers. 2008;**68**(6):1013-1051

[10] Sashiwa H, Shigemasa Y. Chemical modification of chitin and chitosan 2: Preparation and water soluble property of N-acylated or N-alkylated partially deacetylated chitins. Carbohydrate Polymers. 1999;**39**(2):127-138

[11] Lin SC, Buehler MJ. Mechanics and molecular filtration performance of graphyne nanoweb membranes for selective water purification. Nanoscale. 2013;**5**:11801-11807. DOI: 10.1039/C3NR03241H

[12] Kumar Dutta P, Dutta J, Tripathi VS. Chitin and chitosan: Chemistry, properties & applications. Journal of Scientific Industrial Research. 2004;**63**:20-31

[13] Peter MG, Domard A, Muzzarelli RAA. Advances in chitin science. In: Krasavtsev VE, editor. Production and Applications of Chitin and Chitosan Krill as a promising raw material for the production of chitin in Europe. Vol. 4. Potsdam, Germany: Universität Potsdam; 2000. pp. 1. 2000. Chapter -1. ISBN 3-9806494-5-8

[14] Hudson SM, Jenkins DW. Chitin & Chitosan, Encyclopedia of Polymer Science and Technology. 3rd ed. New York: Wiley Interscience

[15] Corine G, Reboul J, Bonneb M, Lebeau B. Ecodesign of ordered mesoporous silica materials. Chemical Society Reviews. 2013;**42**:4217. DOI: 10.1039/c3cs35451b

[16] Dutta PK, Ravikumar MNV, Dutta J. Chitin and chitosan for versatile applications. Journal of Macromolecular Science-Polymer Reviews. 2002;**C42**:307

[17] Bhatia S, editor. Natural Polymer Drug Delivery Systems, Nano- Particles Plant & Algae. Nanoparticles Types, Classification, Characterization, Fabrication Methods and Drug Delivery Applications. pp. 33-93. Chapter 1. DOI: 10.1007/978-3-319-41129-3_2. ISBN 978-3-319-41129-3

[18] Noipa T, Ngamdee K, Tuntulani T, Ngeontae W. Cysteamine CdS quantum dots decorated with Fe^{3+} as a fluorescence sensor for the detection of PPi. Spectrochimica Acta—Part A: Molecular & Biomolecular Spectroscopy. 2014;**118**:17-23

[19] Yang W, Li X, Li Y, Zhu R, Pang H. Applications of Metal-Organic Framework-Derived Carbon Materials. Advanced Materials. Feb 2019;**31**(6): e1804740. DOI: 10.1002/adma.201804740. PMID: 30548705

[20] Yan J-J, Wang H, Zhou Q-H, You YZ. Reversible and multisensitive quantum dot gels. Macromolecules. 2011;**44**(11):4306-4312

[21] Hardison D, Deepthike HU, Senevirathna W, Pathirathne T, Wells MJ. Temperature-sensitive microcapsules with variable optical signatures based on incorporation of quantum dots into a highly biocompatible hydrogel. Material Chemistry. 2008;**18**(44):5368-5375

[22] Shcherban ND, Ilyin VG. Preparation, physicochemical properties and functional characteristics of micromesoporous zeolite materials. Theoretical and Experimental Chemistry. 2016;**51**(6):339-357. DOI: 10.1007/s11237-016-9435-0

[23] Sá-Lima H, Caridade SG, Mano JF, Reis RL. Stimuli-responsive chitosan-starch injectable hydrogels combined with encapsulated adipose-derived stromal cells for articular cartilage regeneration. Soft Matter. 2010;**6**(20):5184-5195

[24] Li X, Zhang H Chi L. On-Surface Synthesis of Graphyne-Based Nanostructures. Advanced Materials. 2018 Dec 28:e1804087. DOI: 10.1002/adma.201804087. PMID: 30592340 DOI: 10.1002/adma.201804087

[25] Lopez-Orozco S, Inayat A, Schwab A, Selvam T, Schwieger W. Zeolitic materials with hierarchical porous structures. Advanced Materials. 2011;**23**(22-23):2602-2615. DOI: 10.1002/adma.201100462

[26] Kayan DB, Koçak D, İlhan M. The activity of PAni-chitosan composite film decorated with Pt nanoparticles for electrocatalytic hydrogen generation. International Journal of Hydrogen Energy. 2016;**41**(25):10522-10529

[27] Kiba S, Suzuki N, Okawauchi Y, Yamauchi Y. Prototype of low thermal expansion materials: Fabrication of mesoporous silica/polymer composites with densely filled polymer inside mesopore space. Chemistry, an Asian Journal. 2010;**5**:2100-2105

[28] Davis ME. Ordered porous materials for emerging applications. Nature. 2002;**417**:813-821. DOI: 10.1038/nature00785